Springer Theses

Recognizing Outstanding Ph.D. Research

Aims and Scope

The series "Springer Theses" brings together a selection of the very best Ph.D. theses from around the world and across the physical sciences. Nominated and endorsed by two recognized specialists, each published volume has been selected for its scientific excellence and the high impact of its contents for the pertinent field of research. For greater accessibility to non-specialists, the published versions include an extended introduction, as well as a foreword by the student's supervisor explaining the special relevance of the work for the field. As a whole, the series will provide a valuable resource both for newcomers to the research fields described, and for other scientists seeking detailed background information on special questions. Finally, it provides an accredited documentation of the valuable contributions made by today's younger generation of scientists.

Theses are accepted into the series by invited nomination only and must fulfill all of the following criteria

- They must be written in good English.
- The topic should fall within the confines of Chemistry, Physics, Earth Sciences, Engineering and related interdisciplinary fields such as Materials, Nanoscience, Chemical Engineering, Complex Systems and Biophysics.
- The work reported in the thesis must represent a significant scientific advance.
- If the thesis includes previously published material, permission to reproduce this must be gained from the respective copyright holder.
- They must have been examined and passed during the 12 months prior to nomination.
- Each thesis should include a foreword by the supervisor outlining the significance of its content.
- The theses should have a clearly defined structure including an introduction accessible to scientists not expert in that particular field.

More information about this series at http://www.springer.com/series/8790

Da Zhao

Functionalization of Carborane via Carboryne Intermediates

Doctoral Thesis accepted by
The Chinese University of Hong Kong, Hong Kong

Springer

Author
Dr. Da Zhao
Department of Chemistry and State Key Laboratory of Synthetic Chemistry
The Chinese University of Hong Kong
Hong Kong
China

Supervisor
Prof. Zuowei Xie
Department of Chemistry
The Chinese University of Hong Kong
Hong Kong
China

ISSN 2190-5053 ISSN 2190-5061 (electronic)
Springer Theses
ISBN 978-981-10-1568-7 ISBN 978-981-10-1569-4 (eBook)
DOI 10.1007/978-981-10-1569-4

Library of Congress Control Number: 2016940355

Printed on acid-free paper

This Springer imprint is published by Springer Nature
The registered company is Springer Science+Business Media Singapore Pte Ltd.

Parts of this thesis have been published in the following journal articles:

1. **Zhao, D.**; Zhang, J.; Xie, Z*. 1,3-Dehydro-*o*-Carborane: Generation and Reaction with Arenes. *Angew. Chem. Int. Ed.* **2014**, *53*, 8488–8491.

2. **Zhao, D.**; Zhang, J.; Xie, Z*. Regioselective Insertion of *o*-Carborynes into the α-C–H Bond of Tertiary Amines: Synthesis of α-Carboranylated Amines. *Angew. Chem. Int. Ed.* **2014**, *53*, 12902–12906.

3. **Zhao, D.**; Zhang, J.; Xie, Z*. Facile Synthesis of *o*-Carborane-Substituted Alkenes and Allenes by a Regioselective Ene Reaction of 1,3-Dehydro-*o*-Carborane. *Chem. –Eur. J.* **2015**, *21*, 10334–10337.

4. **Zhao, D.**; Zhang, J.; Xie, Z*. Dearomative [2 + 2] Cycloaddition and Formal C–H Insertion Reaction of *o*-Carboryne with Indoles: Synthesis of Carborane-Functionalized Heterocycles. *J. Am. Chem. Soc.* **2015**, *137*, 9423–9428.

5. **Zhao, D.**; Zhang, J.; Xie, Z*. An Unprecedented Formal [5 + 2] Cycloaddition of Nitrones with *o*-Carboryne via Tandem [3 + 2] Cycloaddition/Oxygen Migration/Aromatization Sequence. *J. Am. Chem. Soc.* **2015**, *137*, 13938–13942.

6. **Zhao, D.**; Xie, Z*. Recent Advances in the Chemistry of Carborynes. *Coord. Chem. Rev.* **2016**, *314*, 14–33.

7. **Zhao, D.**; Xie, Z*. Visible-Light-Promoted Photocatalytic B–C Coupling via Boron-Centered Carboranyl Radical: Facile Synthesis of B(3)-Arylated *o*-Carboranes. *Angew. Chem. Int. Ed.* **2016**, *55*, 3166–3170.

Supervisor's Foreword

This thesis describes the scientific achievements of Dr. Da Zhao, which were performed during his doctoral program in Department of Chemistry at the Chinese University of Hong Kong. Dr. Zhao has done ground-breaking research on carborane chemistry. Focusing on the area of carborane functionalization, he has done exceptional works by utilizing carboryne intermediates as powerful synthons. As his Ph.D. supervisor, I would like to highlight three important findings of his doctoral work. One is the synthesis of previously inaccessible functional carboranes in a single step from very simple precursors, the second is the first generation of 1,3-dehydro-*o*-carborane featuring a C–B multiple bond, and the third is in situ generation of an electrophilic boron-centered carboranyl radical.

Benzyne has been one of the most studied intermediates in organic synthesis for decades. However, as a three-dimensional relative of benzyne, carboryne intermediate has not enjoyed widespread use in synthesis. Dr. Zhao has found several new reactivity patterns of 1,2-dehydro-*o*-carborane including sp^3 C–H insertion reaction with tertiary amines and [2 + 2]/[5 + 2] cycloaddition reactions with indoles/nitrones, which demonstrate the uniqueness of carboryne intermediates as no similar process has been observed for benzyne.

The known benzyne derivatives contain a formal carbon–carbon triple bond. Dr. Zhao has designed a new precursor, 3-diazonium-*o*-carborane tetrafluoroborate, for producing 1,3-dehydro-*o*-carborane that contains a C–B multiple bond. This reactive species can undergo regioselective [4 + 2] cycloaddition reaction with arenes and ene reaction with alkene/alkynes, which serves as a new methodology for efficient simultaneous functionalization of both cage carbon and boron vertices.

On the other hand, Dr. Zhao has also developed a new method for in situ generation of the boron-centered carboranyl radical. In the presence of a photoredox catalyst eosin Y, 3-diazonium-*o*-carborane tetrafluoroborate can be efficiently converted into the corresponding boron-centered carboranyl radical under visible light irradiation, which can react with a wide range of (hetero)arenes to give a series of carborane-functionalized materials.

The findings in this thesis have been published in top chemistry journals (*Angew. Chem. Int. Ed.* & *J. Am. Chem. Soc.*) and break new ground in carborane chemistry. I hope that these research works could draw more attention towards carborane chemistry and stimulate more researchers to work in this amazing field.

Hong Kong, China
April 2016

Prof. Zuowei Xie

Acknowledgments

I would like to give my sincere thanks to my supervisor, Prof. Zuowei Xie, for giving me the opportunity to work in his research group and study the fascinating chemistry of carborane. Professor Xie is a mentor and friend, from whom I have learnt the vital skill of disciplined critical thinking. I am grateful for his guidance, encouragement, and help during my study at the Chinese University of Hong Kong. Besides my supervisor, I would like to thank Prof. Zhenyang Lin from the Hong Kong University of Science and Technology, for his valuable guidance and very helpful discussion on the theoretical work. His suggestions provide me a better understanding of the chemistry of carboryne intermediates. Additionally, I am very grateful for the friendship of the Lin group.

I am also grateful to my groupmates Dr. Shikuo Ren, Dr. Zaozao Qiu, Dr. Jian Zhang, Dr. Fangrui Zheng, Dr. Sunewang R Wang, Mr. Yang Wang, Mr. Xiao He, Dr. Jiji Zhang, Dr. Xiaoguang Zhou, Dr. Xiong Hu, Dr. Hao Wang, Dr. Jingying Yang, Dr. Li Xiang, Dr. Yangjian Quan, Mr. Tek-Long Chan, Mrs. Jie Zhang, Ms. Cen Tang, and Mr. Yui-Tsz Hin for our happy times inside and outside the chemistry world. I would like to thank Mrs. Hoi-Shan Chan for the single-crystal X-ray analyses, Mrs. Hau-Yan Ng for mass spectra measurement, and Mr. Chun-Wah Lin for GC-MS measurement. I also thank all the staff and technicians in the Department of Chemistry and the graduate school for their help and support during the course of my study.

I am greatly indebted to the Chinese University of Hong Kong and the Hong Kong Research Grants Council for the award of Hong Kong Ph.D. Fellowship and the financial support. It is a pleasure to thank my colleagues and friends at the the Chinese University of Hong Kong, Dr. Chao Cheng, Dr. Shuaijun Yang, Dr. Yafeng Zhang, Dr. Xiaojie Li, Mr. Renxin Mao, Dr. Yun Lei, Dr. Ting Hu, Dr. Yinxue Jin, Dr. Liang Shan, Dr. Liulin Deng, Prof. Qian Miao's group and Prof. Henry N.C. Wong's group for the wonderful times we shared.

I would also thank my wife, Dr. Jiji Zhang, without whose love, encouragement, and discussion, I would not have finished this thesis. Her tolerance of my occasional vulgar moods is a testament in itself of her unyielding devotion and love. Last, but definitely not least, I want to thank my parents for their unconditional love, care, understanding, and support throughout these years.

Contents

List of Figures

List of Tables

Chapter 1
Introduction

Boron compounds, such as borax, have been known and used by ancient cultures for thousand of years. However, element boron itself was not identified until 1808. In contrast to hydrocarbons, boranes avoid the formation of chain structures and clearly prefer the formation of polyhedral clusters [1]. Their structures cannot be described in terms of common organic bond diagrams, in which a connection line between two atoms explicitly represents an electron pair. Instead, connecting lines in their structures only indicate the polyhedral geometry. As shown in Fig. 1.1, typical molecular frameworks of boranes consist of *closo*, *nido*, *arachno*, and *hypho* structures, and the latter three are formally derived from the *closo* frameworks by removal of one, two, or three missing vertices, respectively [2].

As intuitively expressed in its formal nomenclature, carboranes are generally accepted as polyhedral boranes with one or more BH vertices being replaced by CH. The range of composition in carborane cages extends from boron-rich clusters such as $CB_{11}H_{12}^-$ and $C_2B_{10}H_{12}$ to species having as many as six skeletal carbon atoms, but those with high boron content are still dominant. Different from classical organoboranes, the skeletal carbon atoms in carboranes typically have at least three neighbors including hydrogen or other attached substituents in the cluster [1].

Of all known carborane family, *closo*-1,2-dicarbododecaborane(12) or *o*-carborane (1,2-$C_2B_{10}H_{12}$) is the most widely investigated one, mainly due to its commercial availability. Among those synthetic routes, a principal approach to *o*-carborane is the reaction of alkynes with decaborane(12)–Lewis base adducts (Scheme 1.1) [3, 4]. According to the relative position of the two CH vertices, icosahedral carboranes $C_2B_{10}H_{12}$ can exist as three isomers, the *ortho*-, *meta*-, and *para*-isomer (Scheme 1.1). Due to its fascinating features such as spherical geometry and hydrophobic molecular surface and remarkable thermal and chemical stability, carboranes have been found numerous applications in polymers, ceramics, catalysts, and medicals [1, 5]. Besides serving as an ideal source of boron in boron neutron capture therapy (BNCT), newly merging biomedical and other applications of carborane have been recently recognized [6].

D. Zhao, *Functionalization of Carborane via Carboryne Intermediates*,
Springer Theses, DOI 10.1007/978-981-10-1569-4_1

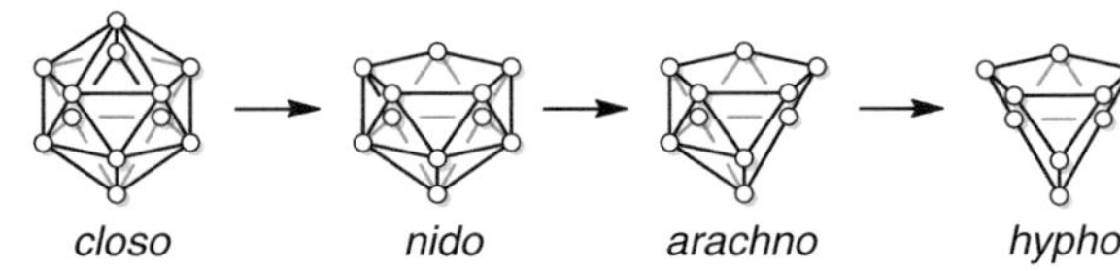

Fig. 1.1 Cage frameworks based on a 12-vertex icosahedral cluster

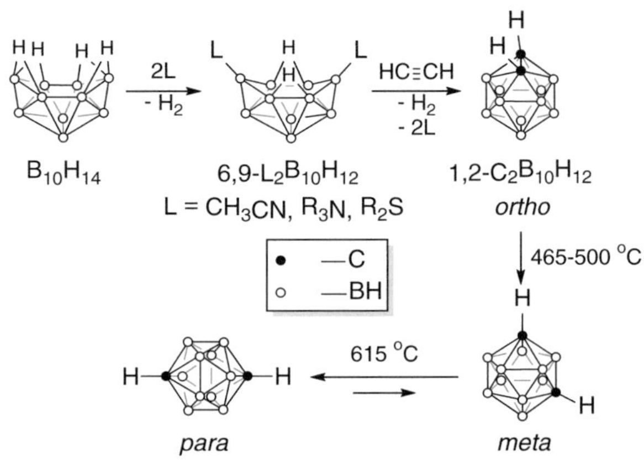

Scheme 1.1 Synthesis of the carborane isomers

1.1 Skeletal Transformation

1.1.1 Thermal Cage Rearrangement

The two remaining isomers, 1, 7- and 1, 12-$C_2B_{10}H_{12}$ (*m*-carborane and *p*-carborane, respectively), can be prepared from thermal cage rearrangement of the *ortho*-isomer [7]. Under inert atmosphere, thermal isomerization of the *o*-carborane in an autoclave first generated the *meta*-isomer one (465–500 °C) and finally the *para*-isomer (615–700 °C) (Scheme 1.1). The former isomerization is irreversible, whereas the latter isomerization is reversible. On the other hand, higher temperatures lead to decomposition. The driving force of the thermal process is rationalized as mutual repulsion between the relatively electropositive (6+) carbon nuclei in the cage.

1.1.2 Deboronation

o-Carborane has been well documented to undergo degradation to the *nido*-7,8-$C_2B_9H_{12}^-$ anion in the presence of strong Lewis base (Scheme 1.2) [8–11]. The mechanism of this degradation process has been studied, and this process that removal of a boron atom from the carborane cluster framework is described

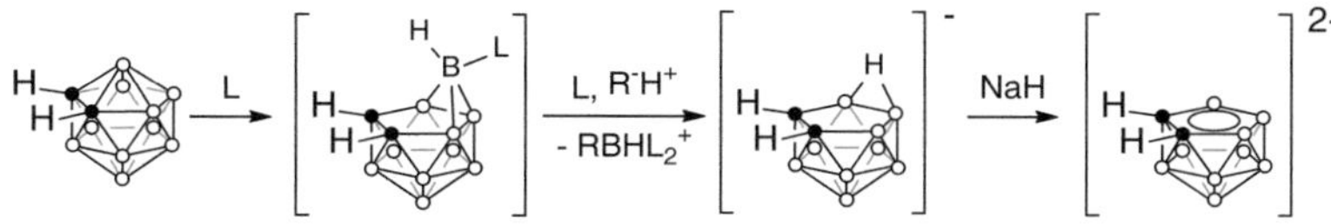

Scheme 1.2 Base-promoted deboronation of *o*-carborane

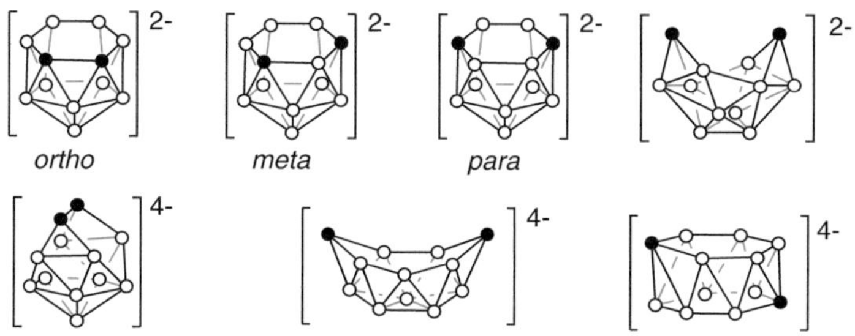

Fig. 1.2 Reported carborane anions of the C_2B_{10} system

as "deboronation." Further deprotonation of the resulting *nido*-$C_2B_9H_{12}^-$ anion with strong base, such as NaH and ^{n}BuLi, gives the corresponding dianion *nido*-$C_2B_9H_{11}^{2-}$, an isolobal analog of cyclopentadienyl anion, $C_5H_5^-$, which makes it very useful in synthesis of metallacarboranes [12, 13] and B(3)-substituted carboranes [14].

1.1.3 Reductive Cage Opening

Due to their electron-deficient nature, carboranes are readily reduced by group 1 metals to form mono-, di-, or tetraanionic species (Fig. 1.2) [15, 16]. Examples of single-electron reduction of carborane are rare [17], whereas two-electron reduction of *o*-$R_2C_2B_{10}H_{10}$ (R=H, alkyl, aryl) with group 1 metals is a well-established process [18]. The reduced species are powerful synthons for metallacarboranes [19] or supercarboranes (carboranes with more than 12 vertices) [20, 21].

Although *o*-carborane cage can undergo skeletal transformations under certain situations, one of the most important features of the carborane system is its ability to enter into substitution reactions at both the carbon and boron atoms without disruption of the cage. These reactions will be discussed in the following two parts: boron substitution and carbon substitution.

1.2 Boron Substitution

1.2.1 Electrophilic Substitution

Carboranes are generally recognized as a three-dimensional inorganic benzene analogue in which the 13 filled bonding MOs are occupied by 26 skeletal electrons. Indeed, they are termed as "superaromatic" [22] and exhibit extraordinary thermal stability as well as unusual chemical reactivity such as electrophilic substitution, similar to that of planar aromatics (Scheme 1.3) [23, 24].

1.2.2 Nucleophilic Substitution

Similar to the aromatic benzene systems, nucleophilic substitution reaction also proceeds well in the case of *o*-carboranes that bearing excellent boron leaving groups, such as N_2^+ and PhI^+ substituents (Scheme 1.4) [25–28].

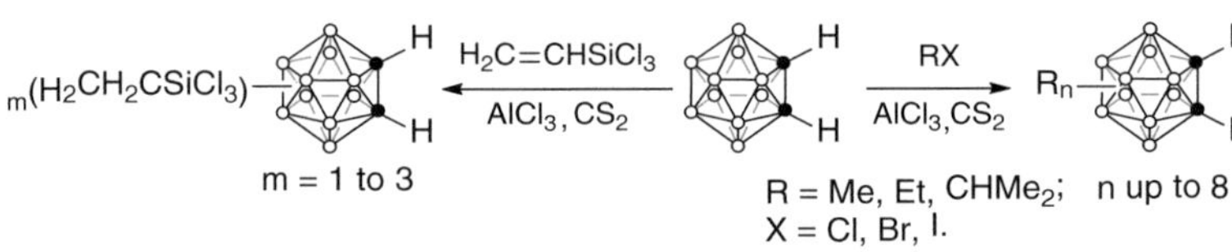

Scheme 1.3 Electrophilic alkylation of *o*-carborane

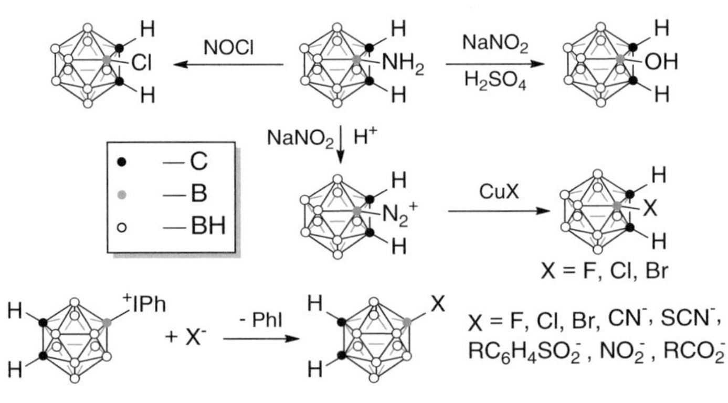

Scheme 1.4 Examples of nucleophilic substitution reaction

1.2.3 Direct B–H Bond Functionalization

There are three feasible mechanistic pathways reported for the direct functionalization of *o*-carborane B–H bonds: electrophilic metalation with metal complexes in high oxidative state such as mercuration [29] and thallation [30]; oxidative addition of a carborane B–H bond to low valent and coordinatively unsaturated transition-metal complexes such as iridium species [31]; and oxidation of the B–H bond by strong oxidants, for example, hydrogen peroxide [32] or nitric acid [33] can oxidize the B–H bond to B–OH or B–N bond, respectively (Scheme 1.5).

1.2.4 Carbene/Carbenoid Insertion of B–H Bond

Carbenes generated by photo irradiation can insert into the B–H bonds of *o*-carboranes to give all four possible isomeric B-substituted monocarboxylic acids, whereas the cage C–H bonds remain unaffected [34, 35]. Jones and coworkers reported an intramolecular version of similar insertion under copper catalysis, in which the resulting product can be further converted to C,B-fused benzocarborane (Scheme 1.6) [36].

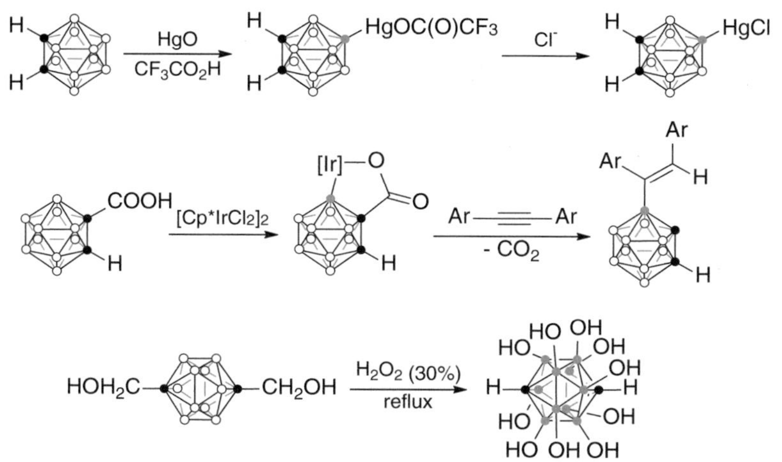

Scheme 1.5 Direct B–H functionalization of *o*-carborane

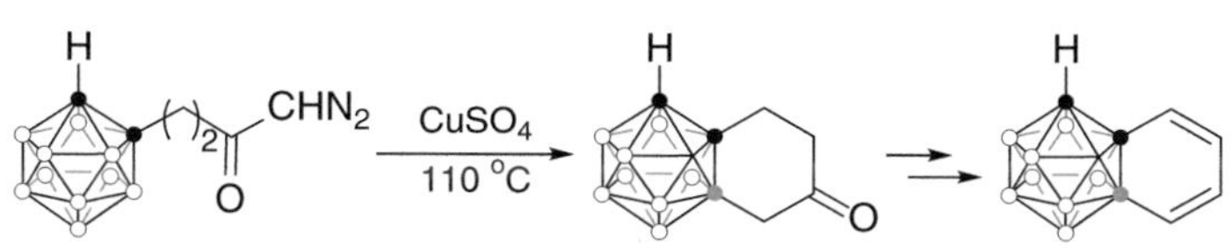

Scheme 1.6 Regioselective coppercarbenoid insertion into B–H bond

1.3 Carbon Substitution

1.3.1 Nucleophilic Substitution

Although various C-monosubstituted and C,C′-disubstituted *o*-carboranes are available from the Lewis base-promoted reactions of suitable alkynes with decaborane as shown in Scheme 1.1, a more popular methodology exploits the high acidity of the cage C–H bond (with the p*K*a value of ~23 for *o*-carborane and ~28 for *m*-carborane) [37], of which one or both can be readily deprotonated by strong bases such as *n*-butyllithium, phenyllithium, or Grignard reagents [38]. Nucleophilic addition between the resulting *o*-carboranyl anions and electrophiles affords various C-substituted *o*-carboranes (Scheme 1.7). Thus, they are remarkably versatile synthons for cage carbon functionalized *o*-carboranes.

Yamamoto and coworkers reported that direct intermolecular and intramolecular addition of *o*-carborane to aldehydes and ketones proceeds smoothly in the presence of tetrabutylammonium fluoride (TBAF), providing a practically useful synthesis of *o*-carboranyl alcohols (Scheme 1.8) [39, 40].

1.3.2 Transmetalation

Salt elimination between lithiocarboranes with metal halides or pseudo-metal halides adds yet another dimension for functionalization on the cage carbon of *o*-carborane and its derivatives. Main group metals [41–43], transition metals [44], as well as pseudo-metal elements [45–49] are all successfully introduced to the cage carbon by this method.

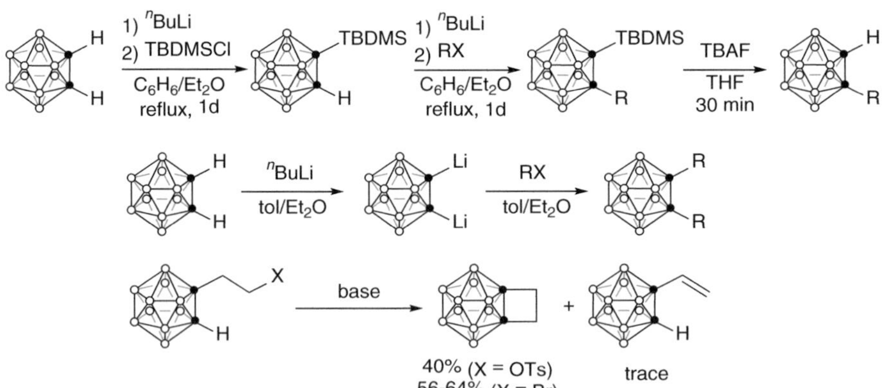

Scheme 1.7 Routes to C-substituted *o*-carboranes

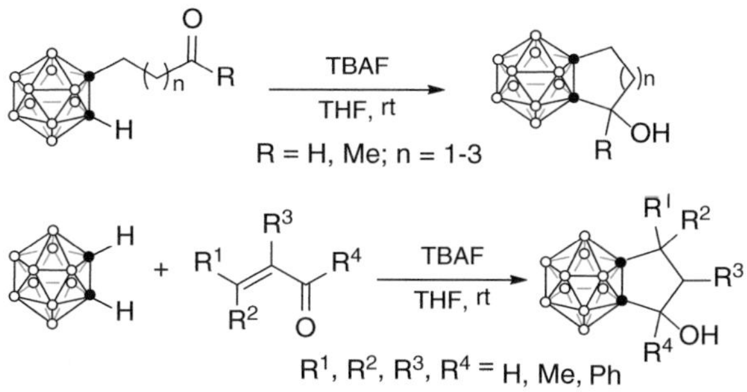

Scheme 1.8 TBAF-promoted synthesis of carboracycles

1.3.3 Lithium–Halogen Exchange

Lithium–halogen exchange between lithiocarboranes and electrophilic halogen species, such as elemental halogens, CCl_4, *N*-fluorobenzenesulfonamide, *p*-toluenesulfonyl chloride, ICl, and $PhI(OAc)_2$ [50–52], leads to the formation of carboranes with cage carbon-halogen bonds. The exchange reaction also shows significant applications in the preparation of precursors for 1,2-dehydro-*o*-carborane (*o*-carboryne), a very reactive intermediate as a three-dimensional analogue to benzyne.

1.4 Dehydrogenation to *o*-Carboryne

One important feature of *o*-carborane is that they can undergo dehydrogenation to form 1,2-dehydro-*o*-carborane (*o*-carboryne) [53], a very reactive intermediates, which can be regarded as a three-dimensional relative of 1,2-dehydrobenzene (benzyne) [53d]. *o*-Carboryne was first reported by Jones in 1990 [53a]. During our exploration, we found that it exists in two resonance forms, a bonding form and a biradical form (Fig. 1.3) [54].

Like the critical role of arynes in modern arene chemistry, *o*-carborynes serve as very useful synthons for functional *o*-carboranes that have found many applications in medicine [6], materials science [5], and organometallic/coordination

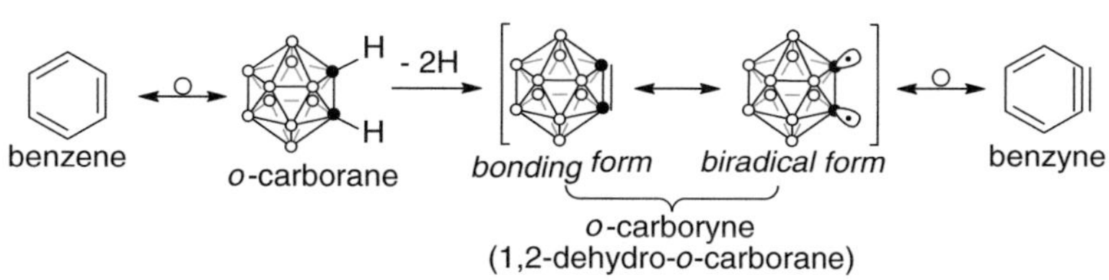

Fig. 1.3 *o*-carborane, *o*-carboryne, and benzyne

chemistry [12]. Many processes, including cycloadditions, ene reactions, and C–H insertion reactions of *o*-carborynes, have been developed to enable the synthesis of a large series of *o*-carboranes derivatives [53d].

1.4.1 Generation and Trapping of o*-Carborynes*

Because of their extreme reactivity, *o*-carborynes must be generated in situ. The reported generation methods are summarized in Scheme 1.9.

1.4.1.1 From 1-Bromo-2-Lithio-*o*-Carborane

In 1990, Jones and coworkers first reported an experimentally feasible precursor, 1-bromo-2-lithio-*o*-carborane (**A**), by the treatment of 1,2-dilithio-*o*-carborane with one equivalent of bromine in diethyl ether. Different from 1-Br-2-Li-C_6H_4, an earlier precursor of benzyne, which requires temperatures as low as −100 °C to avoid rapid loss of LiBr, this precursor is stable at 0 °C and heating above room temperature is important for the production of *o*-carboryne (Scheme 1.9, method a) [53].

The resulting *o*-carboryne intermediate was successfully trapped by a series substrate in percyclic or ene reaction fashions, which provided the definitive evidence for the existence of such intermediate (Scheme 1.10) [55]. In the presence of furans or dienes as the trapping agents, products of [4 + 2]/[2 + 2] cycloadducts or ene reaction products were isolated in good yields, similar to that of benzyne [56].

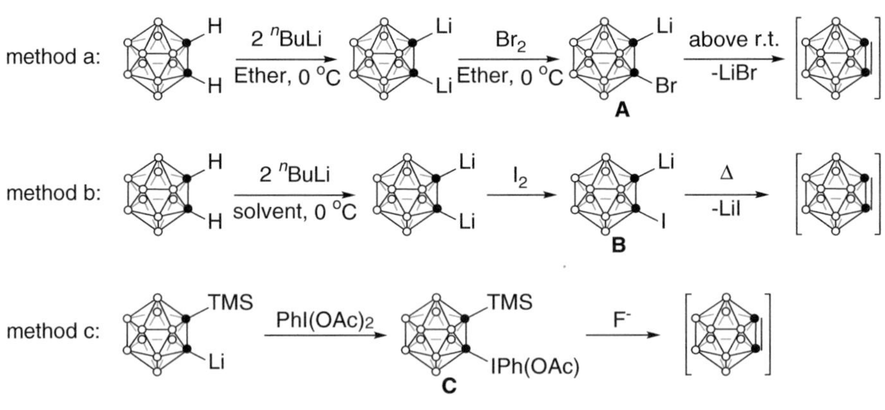

Scheme 1.9 Reported methods for the generation of *o*-carboryne

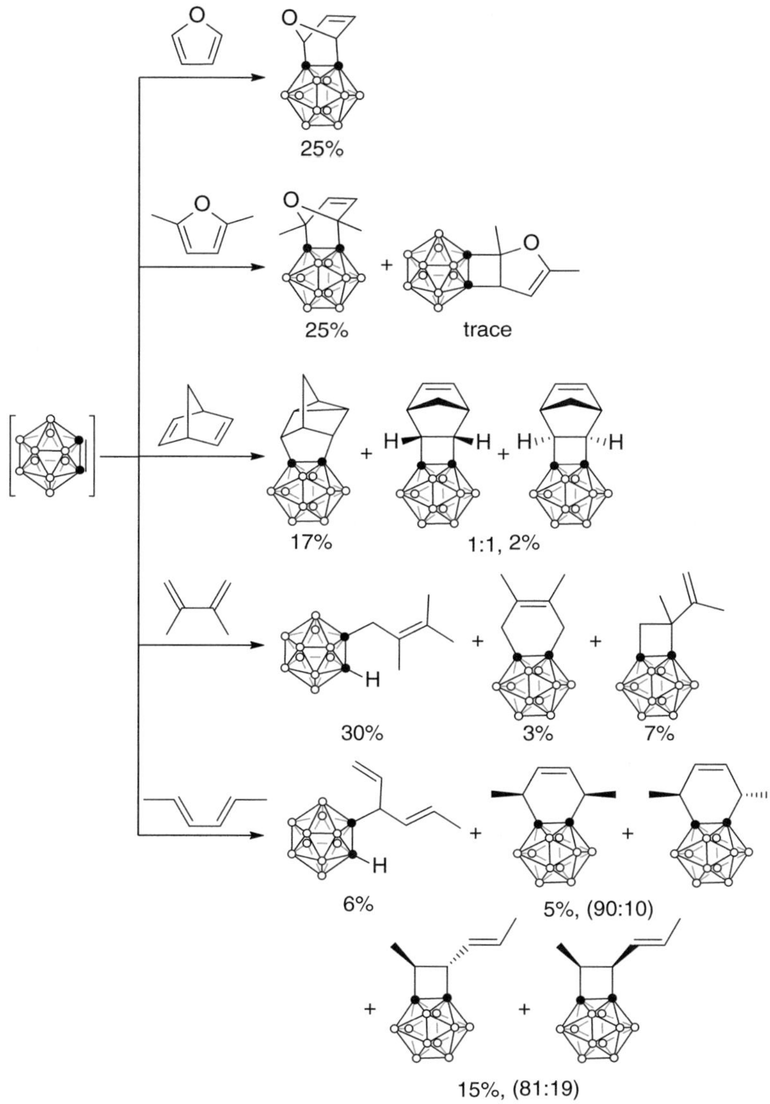

Scheme 1.10 Trapping reactions of *o*-carboryne

1.4.1.2 From 1-Iodo-2-Lithio-*o*-Carborane

In 2010, we have developed another precursor, 1-iodo-2-lithio-*o*-carborane (**B**), for the generation of *o*-carboryne [53c]. It is prepared by the treatment of dilithio-*o*-carborane with a single equiv of solid iodine at room temperature in common organic solvents (Scheme 1.9, method b). It is stable at temperatures below 60 °C, and comparing to method a, the purification and handling of iodine is much easier than that of bromine.

1.4.1.3 From Phenyl[o-(Trimethylsilyl)Carboranyl]Iodonium Acetate

The aforementioned two precursors are strong bases. Kang and coworkers report a neutral precursor, phenyl[*o*-(trimethylsilyl)carboranyl]iodonium acetate (C), which is prepared by the reaction of [*o*-(trimethylsilyl)carboranyl]lithium with $IPh(OAc)_2$ (Scheme 1.9, method c) [53b]. *o*-Carboryne can be produced in the presence of desilylating reagents such as CsF or KF/18-crown-6 [57].

1.4.2 Reactivity

Even at low temperatures, *o*-carborynes are extraordinary reactive. Their reactions can be divided into two groups: the pericyclic reactions and C–H bond insertion reactions. The pericyclic reactions consist of several categories such as Diels–Alder reactions, [2 + 2] cycloadditions, and ene reactions. More recently, the C–H bond insertion reactions of *o*-carborynes have been studied. This new methodology, which has not been reported in the reaction of arynes [58], serves as another efficient approach for various functional *o*-carboranes.

1.4.2.1 Diels–Alder Reactions of *o*-Carboryne

Diels–Alder reaction is one of the most widely studied reactions of *o*-carborynes and is used as both a means of trapping *o*-carborynes (Scheme 1.10) and as a synthetic tool. Due to the highly electrophilic character of *o*-carborynes, the Diels–Alder reaction is observed with a wide range of dienes including simple benzene derivatives as well as many benzenoid aromatic compounds.

Reaction of *o*-carboryne with benzene and polycyclic aromatics gives [4 + 2] cycloaddition products. As suggested in Table 1.1, precursor **B** is a more efficient precursor for *o*-carboryne compared to **A** or **C** as much improved yields are obtained [53a, b, 59, 60]. For the reaction of precursor **B** with anthracene, a minor 1,4-cycloadduct is also isolated in 4 % yield in addition to the expected 9,10-addition product [60], which is not observed in the reactions of **A**, **C**, or benzyne [61]. If the activated central aromatic ring is sterically inaccessible, such as phenanthrene, the [4 + 2] cycloaddition on the less reactive yet more accessible aromatic ring proceeds (Table 1.1, entry 4).

Along with two expected [4 + 2] cycloadducts, several other products are also isolated from the reaction of *o*-carboryne with toluene. After careful analyses, they are identified as one tandem [4 + 2] cycloaddition/homo-Diels–Alder reaction product, one benzylic C–H insertion product, one tandem ene reaction/[2 + 2]

Table 1.1 Reaction of *o*-carboryne with benzene and polycyclic aromatics

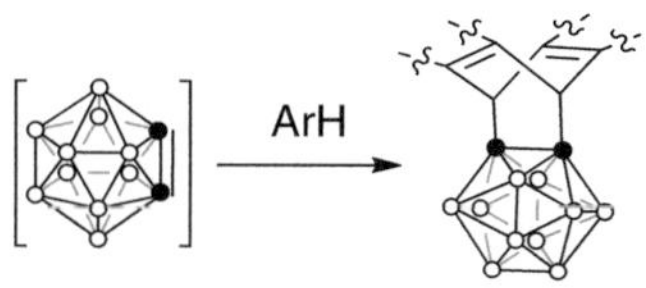

Entry	ArH	Product	Yield (%) using different precursors[a]		
			A	B	C
1	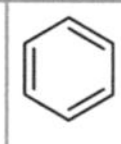	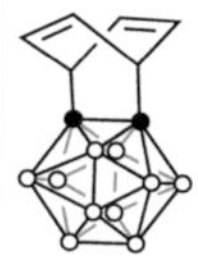	8	25	–[b]
2	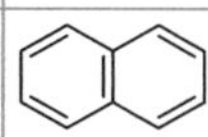	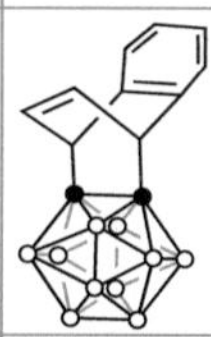	15	60	33
3	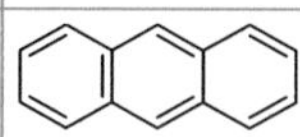	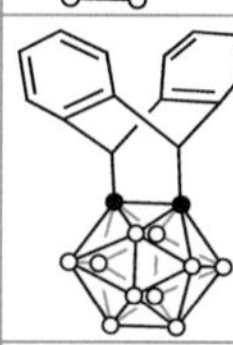	19	64	48
4	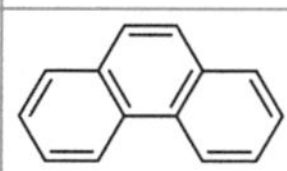	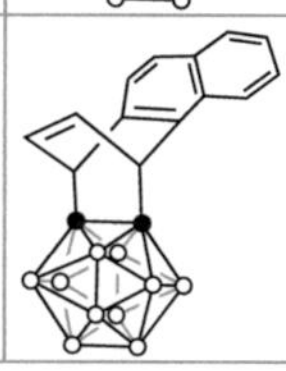	–[b]	25	–[b]

[a]Isolated yields
[b]Data not available

cycloaddition product, and one tandem ene reaction/dehydrogenation product (Scheme 1.11) [60].

The reactions of *o*-carboryne with various alkylbenzenes and trimethylsilylbenzene are also investigated to afford two [4 + 2] cycloaddition isomers as the major products (Scheme 1.12) [60].

As shown in Scheme 1.13, there are two major steric repulsions when *o*-carboryne approaches the aromatic ring. One results from the interactions between the *ipso* carbon and the "fatty" cage in 1,4-addition fashion, and the other comes from the steric repulsions between the substituent R and the flanking cage B(3,6)-H hydrogen atoms in 2,5-addition fashion [55b]. In general, the relative reactivities of

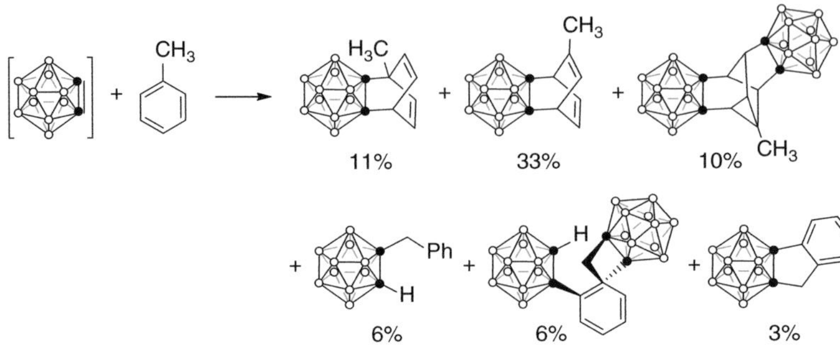

Scheme 1.11 Reaction of *o*-carboryne with toluene

Scheme 1.12 Diels–Alder reaction of *o*-carboryne with alkylbenzenes

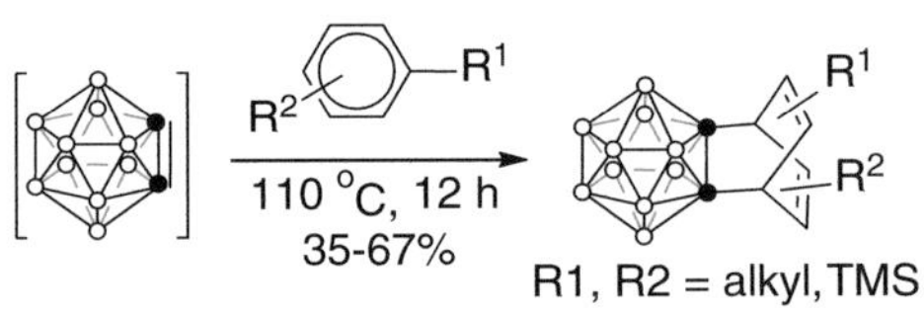

Scheme 1.13 Proposed models for steric repulsions between *o*-carboryne and substituted benzenes

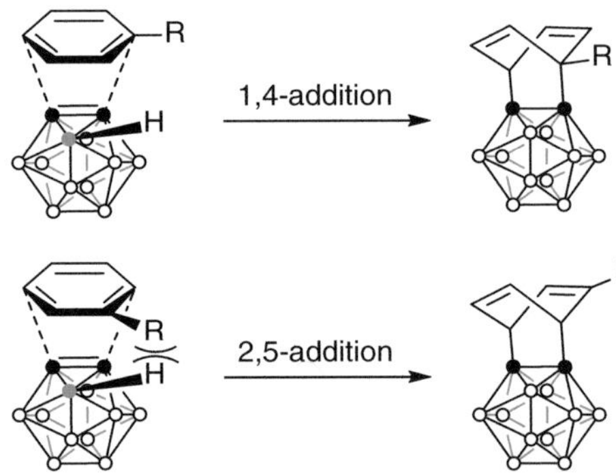

1,4- over 2,5-addition lie around 1, which indicates that the differences between the two types of steric interactions are small.

On the other hand, remarkable differences are observed in the reactions of *tert*-butylbenzene and *p*-xylene as suggested by the ratio of 1,4- over 2,5-adduct. It is noted that the reaction of benzyne with toluene or *tert*-butylbenzene shows a similar regioselectivity [62]. Thus, the interactions between the bulky *tert*-butyl group and the axially flanking hydrogen atom in 2,5-addition fashion are seemingly very strong, which results in the unusual regioselectivity observed in the reaction of *o*-carboryne with *tert*-butylbenzene. Such phenomenon is not observed in the reaction of *o*-carboryne with trimethylsilylbenzene, which can be ascribed to a much longer C–Si bond, reducing significantly steric effects.

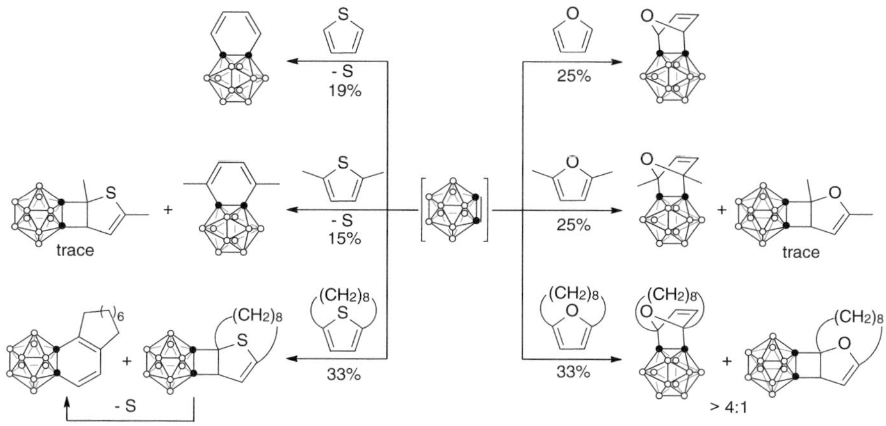

Scheme 1.14 Cycloaddition reaction of *o*-carboryne with furans and thiophenes

The reaction of *o*-carboryne with furan results exclusively in [4 + 2] cycloaddition fashion, and it has been used as a trapping method of *o*-carboryne in the early 1990s [55a]. However, minor [2 + 2] cycloaddition products are also formed in the reactions of *o*-carboryne with 2,5-dimethylfuran and [8] (2,5)-furanophane (Scheme 1.14) [63]. The dominance of [4 + 2] cycloaddition retains in the reaction of *o*-carboryne with thiophene and 2,5-dimethylthiophene and paralleling with the benzyne reaction with thiophenes. The [4 + 2] cycloaddition products in the reaction of *o*-carboryne with thiophenes also undergo desulfurization reaction to afford benzocarboranes. However, reaction with [8] (2,5)-thiophenophane surprisingly proceeds in [2 + 2] fashion (Scheme 1.14) [63].

According to the theoretical calculations, the unusual chemoselectivity observed in the reaction of thiophenophane may result from the extremely higher thermodynamic preference, as a result of the sharp strain energy increase brought on by [4 + 2] addition and the [2 + 2] cycloadduct's relaxed bridge geometry. And the different effect of the octamethylene bridge on the cycloadditions of [8] (2,5)-furanophane and [8] (2,5)-thiophenophane possibly lies in the strain energy changes that accompany the cycloadditions. This difference can be ultimately related to the geometrical differences in the products because of the longer C–S versus C–O bonds and the larger size of S versus O [63].

o-Carboryne reacts with simple styrene derivatives to give extra-annular [4 + 2] cycloadduct as major product although minor [2 + 2] cycloaddition reaction/ene reaction also take place, dependent upon the substituents on the exocyclic double bond (Scheme 1.15) [64].

The resulting [4 + 2] cycloaddition intermediates are so reactive that they undergo rearomatization via either a 1,3-hydrogen rearrangement or dehydrogenation to give 3,4-dihydronaphtho[1, 2]-*o*-carboranes or naphtho[1,2]-*o*-carboranes, respectively. In sharp contrast to that of benzyne [65], further additions of *o*-carboryne onto the primary [4 + 2] cycloadducts are not observed. Substituents

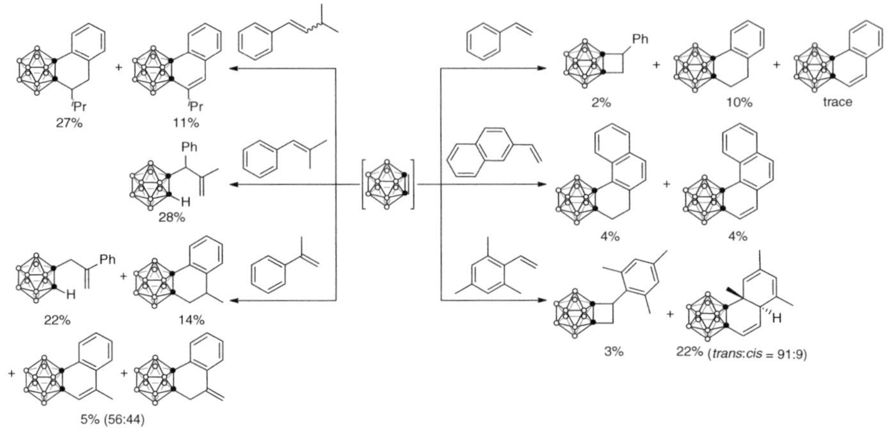

Scheme 1.15 Reaction of *o*-carboryne with styrenes

on C=C double bond can block the cycloaddition. These results may suggest that steric factors play an important role in the chemoselectivity of the [4 + 2] cycloaddition and ene reaction.

1.4.2.2 Ene Reactions of *o*-Carborynes

o-Carboryne can undergo ene reaction with a variety of alkenes, and the mechanistic studies suggest that such ene reactions are likely to be concerted processes [55a, c].

Similar to benzyne [66], *o*-carboryne reacts with simple alkenes exclusively in ene reaction pattern (Scheme 1.16) [55c]. It is noteworthy that in the reaction of benzyne with enamines [67], comparable amount of [2 + 2] cycloadducts are also formed in addition to the ene reaction products, whereas only ene adducts are isolated for *o*-carboryne [55b]. Such preference can be rationalized in terms of the different steric demands of the two intermediates. More importantly, in the presence of *β*-pinene as trapping agent, the absence of any rearrangements in the final product strongly supports that the ene reaction of *o*-carboryne proceeds in a concerted manner [55c]. However, for the reactions of *o*-carboryne with dienes, the chemoselectivity is poor [55]. In addition to the ene reaction, Diels–Alder reaction and [2 + 2] cycloaddition are also observed (Scheme 1.10).

1.4.2.3 [2 + 2] Cycloaddition Reactions of *o*-Carborynes

Although minor [2 + 2] cycloaddition products were observed in the aforementioned reaction of *o*-carboryne with thiophenes, alkenes, and styrenes, examples of

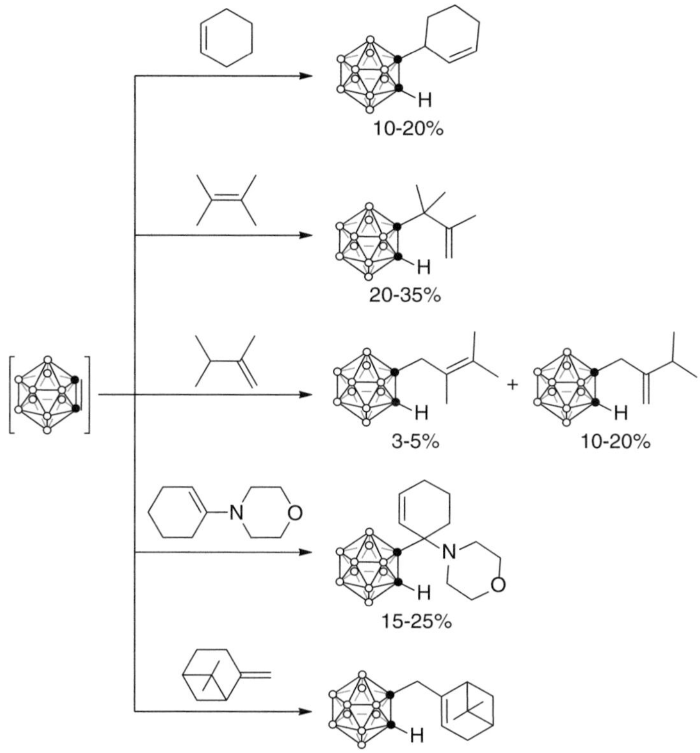

Scheme 1.16 Reaction of *o*-carboryne with aliphatic alkenes

[2 + 2] cycloaddition of *o*-carboryne are still limited, especially comparing to that of benzyne [58].

o-Carboryne reacts with strained cycloalkynes such as cyclooctyne and 3,3,6,6-tetramethyl-1-thia-4-cycloheptyne in exclusive [2 + 2] fashion to give cyclobutenocarboranes [67]. On the other hand, minor ene product is also isolated from 1-phenyl-1-propyne and such ene reaction is dominant in the reaction of aliphatic internal alkynes such as 3-hexyne (Scheme 1.17) [68]. Terminal alkynes are not suitable substrates as they would neutralize the cage C–Li bond of the precursor.

As mentioned earlier, it is documented that *o*-carboryne can undergo [4 + 2] cycloaddtion with alkylbenzenes. When anisoles are examined in these reactions, the formal cycloinsertion products, cyclooctatetraenocarboranes, are isolated as the major products in addition to the minor [4 + 2] cycloaddition products (Scheme 1.18) [69]. It is noted that benzyne reacts with anisoles to generate exclusively the [4 + 2] cycloaddition products, which can be further transformed to benzocyclooctatetraene by photolysis followed with gas phase thermolysis process [70].

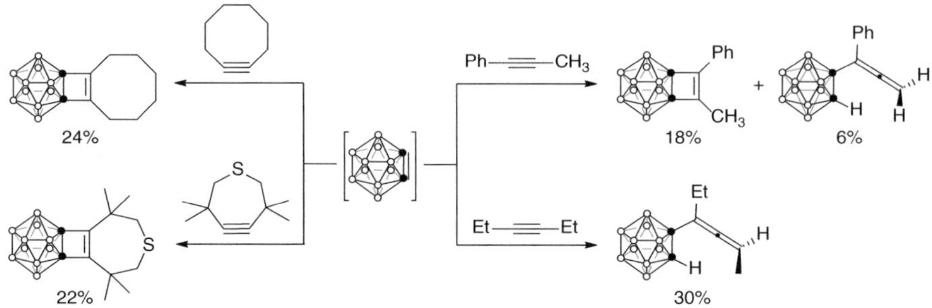

Scheme 1.17 Reaction of *o*-carboryne with alkynes

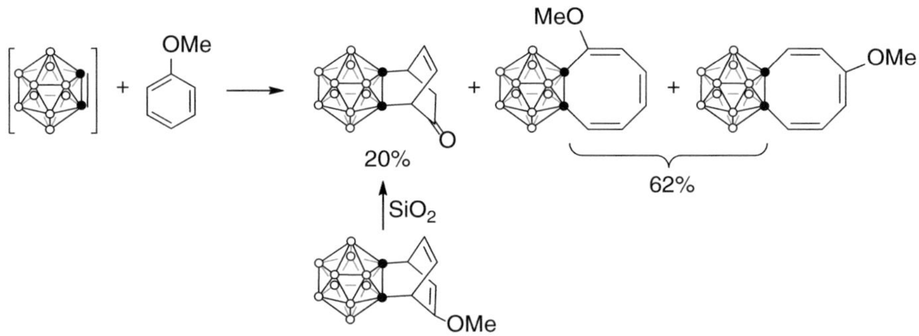

Scheme 1.18 Reaction of *o*-carboryne with anisole

An intermediate, which is derived from the [2 + 2] cycloaddition, is isolated and structurally characterized from the reaction of *o*-carboryne with 2-methylanisole. It can be further converted into the formal cycloinsertion product under standard reaction conditions. On the basis of the experimental results, a plausible mechanism is proposed in Scheme 1.19. *o*-Carboryne can undergo both [4 + 2] and [2 + 2] cycloaddition reactions with anisoles to give two types of dearomatization products.

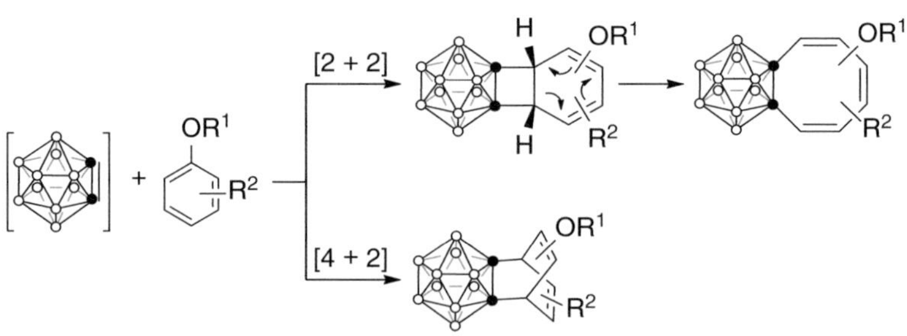

Scheme 1.19 Possible pathways for reaction of *o*-carboryne with anisoles

Thermal [3, 3] sigmatropic rearrangement of the [2 + 2] cycloadducts generates the formal cycloinsertion compounds. Detailed studies also show that steric factors play a role in both regio- and chemoselectivity, and electronic factors have a dramatic influence on the chemoselectivity as only the [4 + 2] cycloaddition was observed with toluene [69].

1.4.2.4 C–H Bond Insertion Reactions of *o*-Carborynes

Experimental data suggest that there is a correspondence in the reactions of *o*-carboryne with benzyne [58]. For instance, the reactivity patterns of *o*-carboryne discussed above, the pericyclic reactions, are also found in the reactions of benzyne. However, during our exploration, we have found that *o*-carboryne has unique properties of its own owing to its steric/electronic features. One example is that it can undergo sp^2/sp^3 C–H bond insertion reactions, which has not been observed for benzyne.

As mentioned in Sect. 1.4.2.3, *o*-carboryne can insert into the C–C bond of the aromactic ring in anisole, which is a semiaromatic ether. When alkyl ethers are subjected to this reaction under UV irradiation (365 or 254 nm), the formal sp^3 C–H bond insertion products are afforded with regioselectivity for secondary C–H bonds being much greater than that of primary C–H bonds [71]. Various cage B-substituents such as phenyl, halo, and methyl groups are well tolerated in this reaction, and electronic factor plays an important role in both regioselectivity and product yields. The experimental data suggest that *o*-carboryne favors insertion into the ethereal -C–H bond, which is different from the C–O bond insertion of benzyne (Scheme 1.20) [72].

Such C–H bond insertion reaction is completely suppressed by a radical scavenger, 2,2,6,6-tetramethylpiperidine-1-oxyl (TEMPO), indicating that a radical process might be involved. On the other hand, the intermolecular primary kinetic isotope effect (KIE) experiments exhibit a significant isotope effect with the k_H/k_D value of 5.3, which indicates that the C–H bond cleavage is the rate-determining step [73].

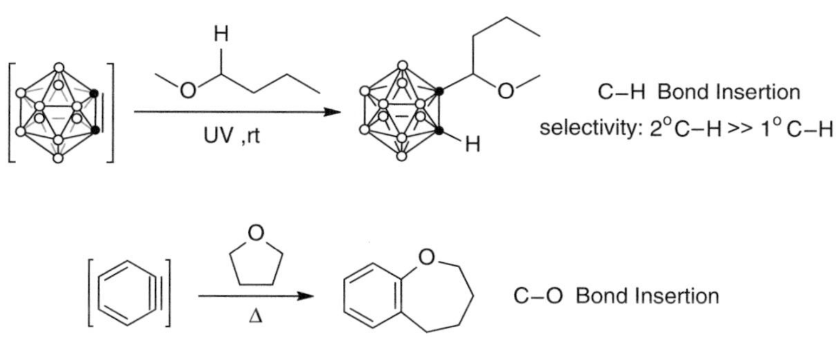

Scheme 1.20 Reaction of ethers with *o*-carboryne and benzyne

Scheme 1.21 Possible pathway for α-carboranylation of ether

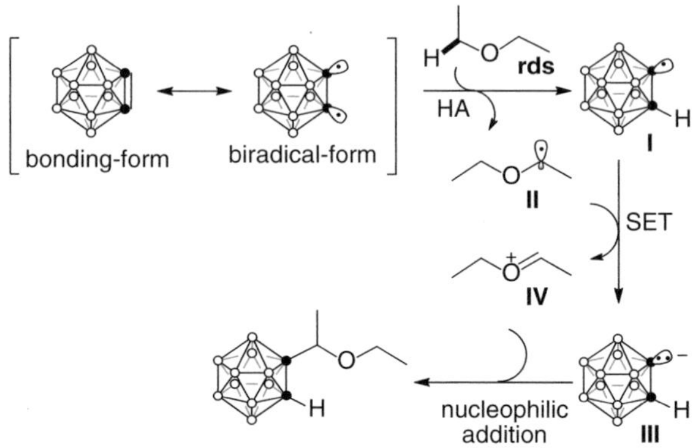

A plausible mechanism is thus proposed in Scheme 1.21. UV irradiation promotes the elimination of LiI from the precursor to form the intermediate *o*-carboryne that is best described as a resonance hybrid of both bonding and biradical forms. It undergoes hydrogen abstraction (HA) with ether to generate radical species **I** and **II**, which is the rate-determining step (RDS). A following fast single-electron transfer (SET) between **I** and **II** produces a carboranyl anion **III** and an electrophile **IV**. Finally, nucleophilic addition of **III** to **IV** yields the final insertion product [71].

In addition to the above sp^3 C–H bond insertion, *o*-carboryne can also undergo sp^2 C–H bond insertion reaction with ferrocene, an intriguing organometallic aromatic species (Scheme 1.22) [74]. This methodology provides a simple access to ferrocenyl (Fc)-substituted *o*-carboranes [75]. It is noted that no C–H bond insertion is observed between benzyne and ferrocene. Various cage B-substituents such as phenyl, halo, and methyl groups are well tolerated in this reaction. In addition to the monoferrocenyl-*o*-carboranes, reaction of ferrocene with 3-halocarborynes also generates bis(ferrocenyl)-*o*-carboranes. It is interesting to note that disubstituted *o*-carboranes are also produced from the C–H bond insertion reaction of *o*-carboryne with aliphatic ethers [71], which may indicate that both reactions share some common features in the reaction mechanism.

Mechanistically, it is suggested that the oxidation of ferrocene by *o*-carboryne to form the electrophilic ferrocenium cation via a single-electron transfer (SET) initiates this insertion reaction. This is supported by the fact that the treatment of 1,2-dilithio-*o*-carborane with ferrocenium hexafluorophosphate also generates monoferrocenyl-*o*-carboranes. A plausible mechanism is proposed in Scheme 1.23 [75]. Nucleophilic

Scheme 1.22 Reaction of *o*-carborynes with ferrocene

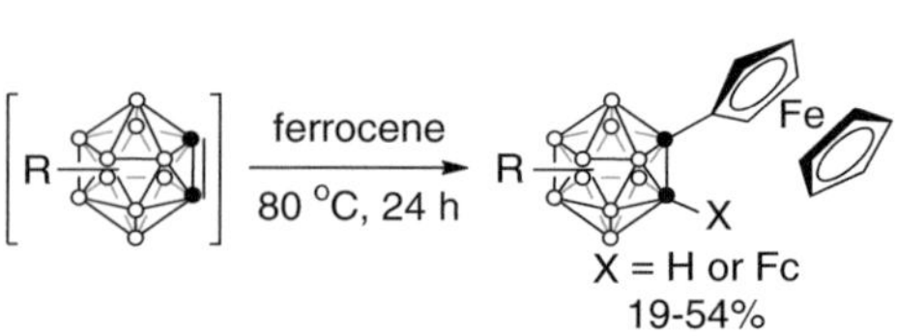

Scheme 1.23 Possible mechanism of *o*-carboryne insertion into ferrocenyl C–H bond

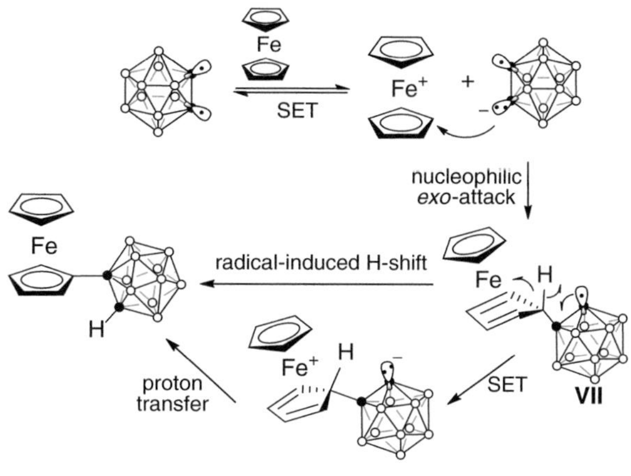

attack of the carboryne radical anion on the cyclopentadienyl ring produces adduct **VII**. Direct radical-induced hydrogen shift of **VII** or intramolecular single-electron transfer followed by an intramolecular protonation gives final C–H bond insertion product.

1.4.3 Transition Metal–Carboryne Complexes

Coordination of *o*-carboryne to transition metals forms relatively stable complexes that featuring a constrained metallacyclopropane, which opens up new opportunities for carboryne chemistry [76]. Such transition metal–carboryne complexes are prepared by transmetalation between 1,2-$Li_2C_2B_{10}H_{10}$ and L_nMCl_2 (L = Ligands, M = Ni, Pd, Pt, Zr), and they are best described as a resonance hybrid of the M–C σ and M–C π bonds (Scheme 1.24) [77, 78].

The nickel–carboryne complex (η^2-$C_2B_{10}H_{10}$)$Ni(PPh_3)_2$ can (i) undergo regioselective [2 + 2 + 2] cycloaddition reactions with two equiv of alkyne to afford benzocarboranes [79a], (ii) react with one equiv of alkene to generate alkenylcarborane-coupling products [79b], and (iii) also undergo a three-component [2 + 2 + 2] cyclotrimerization with one equiv of activated alkene and one equiv of alkyne to give dihydrobenzocarboranes (Scheme 1.25) [79c]. The reaction of *o*-carboryne with alkynes can also proceed in a catalytic way by using nickel catalysts [80].

Scheme 1.24 A general route to transition metal–carboryne complexes

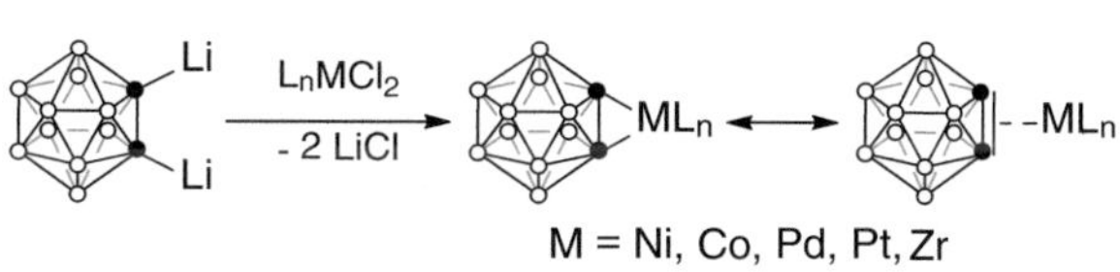

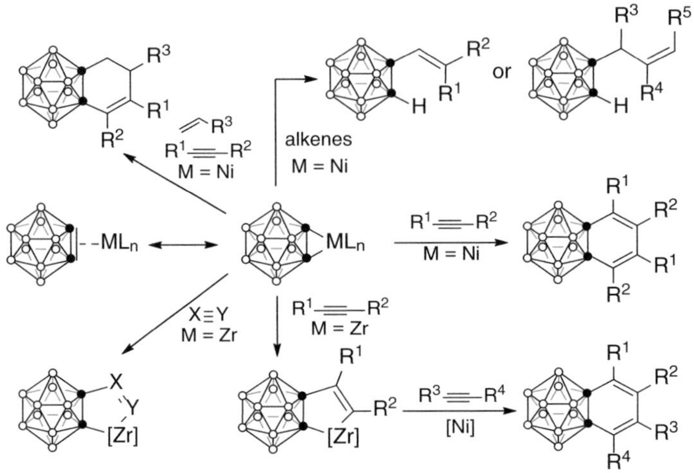

Scheme 1.25 Reactivities of transition metal–carboryne complexes

On the other hand, the zirconium–carboryne species, generated in situ from $Cp_2Zr(\mu\text{-}Cl)(\mu\text{-}C_2B_{10}H_{10})Li(OEt_2)_2$, reacts with only one equiv of alkyne or polar unsaturated organic substrates to give monoinsertion metallacycles, even in the presence of excess substrates (Scheme 1.25) [81]. The resultant five-membered zirconacyclopentenes are an important class of intermediates for the synthesis of a variety of functionalized carboranes. Transmetalation of zirconacyclopentenes to other metals, such as Ni and Cu, was also found to be a very useful tool for various chemical transformations [82].

1.4.4 Bonding and Structure

After a detailed explorations on the reaction between *o*-carboryne with dienes, Jones and coworkers finally reached a conclusion that the *o*-carboryne intermediate *is best formulated as a "normal" dehydro compound, possessing a symmetrical HOMO and unsymmetrical LUMO* [55a]. The LUMO energies for *o*-carboryne, benzyne, and ethylene are calculated as −0.1438, −0.0702, and 0.0188 au, respectively; that is, the difference in energy between the LUMO of *o*-carboryne and the HOMO of butadiene, $E(\text{LUMO}_{\text{ene}}\text{-HOMO}_{\text{diene}})$, is lower than that for ethylene and benzyne [83]. Therefore, in Diels–Alder reactions, the reactivity toward normal dienes is expected to be in the order *o*-carboryne > benzyne > C_2H_4, if steric considerations are not taken into account. The calculation shows that *o*-carboryne is energetically comparable to that of benzyne with the value of *ca.* −99 kcal/mol. The C–C bond length of *o*-carboryne is calculated to be 1.356 Å, which is much shorter than that of 1.625 Å in the parent *o*-carborane, indicating the multiple bond character. As pointed out by Jones and coworkers, *Questions of "doubleness" or "tripleness" are not productive and in this case are essentially meaningless, as the carborane framework is constructed from three-center, two-electron bonding.*[1] Therefore, the remarkably longer bond distance than that calculated for benzyne (1.245 Å) in the same calculation is understandable [83].

1.5 Research Objectives

Efforts of almost three decades in the research area of *o*-carboryne have demonstrated that *o*-carborynes can serve as very useful synthons for generating a variety of functionalized *o*-carborane derivatives that have potential applications in medicine, material science, and organometallic/coordination chemistry. However, in comparison with benzyne, studies of *o*-carborynes are still in the nascent stage. New precursors are desired, and new reactivity patterns need to be explored. Dipolar reactions involving *o*-carborynes still remain unrevealed. Direct construction of *o*-carborane-fused heterocycles via *o*-carboryne intermediates is also unaddressed. How to generate 1,3-dehydo-*o*-carborane, which featuring a cage C–B multiple bond, is another open question in this field [84].

To take these challenges, the objectives of this research are (1) exploration of new reactivities of *o*-carborynes and (2) design and synthesis of precursors for the production and reactivity study of 1,3-dehydo-*o*-carborane. In the following chapters of this thesis, we would like to discuss the details of our work on theses.

References

1. Grimes RN (2011) Carboranes, 2nd edn. Academic Press, Amsterdam
2. Williams RE (1992) The polyborane, carborane, carbocation continuum: architectural patterns. Chem Rev 92:177–207
3. (a) Heying TL, Ager JW, Clark SL et al (1963) A new series of organoboranes. I. Carboranes from reaction of decaborane with acetylenic compounds. Inorg Chem 2(6):1089–1092 (b) Fein MM, Mayes N, Cohen MS et al (1963) Carboranes. II. Preparation of 1- and 1,2-substituted carboranes. Inorg Chem 2(6):1111–1115
4. Tietze LF, Griesbach U, Elsner O (2002) Efficient synthesis of ortho-carborane by reaction of propiolic acid tert-butyl ester with decaborane(14). Synlett 7:1109–1110
5. (a) Hosmane NS (2011) Boron science: new technologies and applications. CRC Press, Boca Raton (b) Grimes RN (2015) Carboranes in the chemist's toolbox. Dalton Trans 44(13):5939–5956 (c) Bregadze VI (1992) Dicarba-closo-dodecaboranes $C_2B_{10}H_{12}$ and their derivatives. Chem Rev 92(2):209–223
6. (a) Issa F, Kassiou M, Rendina LM (2011) Boron in drug discovery: carboranes as unique pharmacophores in biologically active compounds. Chem Rev 111(9):5701–5722

[1] A note made in reference [5] in the reference [55].

(b) Scholz M, Hey-Hawkins E (2011) Carbaboranes as pharmacophores: properties, synthesis, and application strategies. Chem Rev 111(11):7035–7062

7. (a) Edvenson GM, Gaines DF (1990) Thermal-isomerization of regiospecifically B-10-labeled icosahedral carboranes. Inorg Chem 29(6):1210–1216 (b) Grafstein D, Dvorak J (1963) Neoca]boranes, a new family of stable organoboranes isomeric with carboranes. Inorg Chem 2 (6):1128–1133
8. (a) Wiesboeck RA, Hawthorne MF (1964) Dicarbaundecaborane(13) + derivatives. J Am Chem Soc 86(8):1642–1643 (b) Hawthorn MF, Young DC, Garrett PM et al (1968) Preparation and characterization of (3)-1,2- and (3)-1,7-dicarbadodecahydroundecaborate(-1) ions. J Am Chem Soc 90(4):862–868
9. (a) Zakharki Li, Kalinin VN (1965) On reaction of amines with barenes. Tetrahedron Lett 7:407–409 (b) Hawthorn Mf, Wegner PA, Stafford RC (1965) Comments on reaction of amines with 1,2-dicarbaclovododecaborane(12). Inorg Chem 4(11):1675–1675
10. (a) Tomita H, Luu H, Onak T (1991) Cage opening of parent closo cage carboranes with fluoride-ion - formation of [5-f-nido-2,4-c2b4h6]-. Inorg Chem 30(4):812–815 (b) Fox MA, Gill WR, Herbertson PL et al (1996) Deboronation of C-substituted ortho- and meta-closo-carboranes using “wet” fluoride ion solutions. Polyhedron 15(4):565–571 (c) Getman TD (1998) Investigation of potassium fluoride supported on alumina in the deboronation of o-carborane. Inorg Chem 37(13):3422–3423 (d) Yoo J, Hwang JW, Do Y (2001) Facile and mild deboronation of o-carboranes using cesium fluoride. Inorg Chem 40 (3):568–570
11. Willans CE, Kilner CA, Fox MA (2010) Deboronation and deprotonation of ortho-carborane with n-heterocyclic carbenes. Chem-Eur J 16(35):10644–10648
12. (a) Hawthorne MF (1975) Perspectives in metallocarborane chemistry. J Organomet Chem 100(1):97–110 (b) Xie Z (2002) Advances in the chemistry of metallacarboranes of f-block elements. Coord Chem Rev 231(1–2):23–46 (c) Hosmane NS, Maguire JA (2007) d- and f-Block metallacarbaboranes. In: Crabtree RH, Mingos DMP (eds) Comprehensive organometallic chemistry III, Elsevier, Oxford
13. (a) Hawthorn MF (1968) Chemistry of polyhedral species derived from transition metals and carboranes. Acc Chem Res 1(9):281–288 (b) Hawthorn MF, Dunks GB (1972) Metallocarboranes that exhibit novel chemical features—virtually unlimited variety of structural and dynamic features are observed in metallocarborane chemistry. Science 178 (4060):462
14. (a) Hawthorn MF, Wegner PA (1965) B-phenyl-1,2-dicarbaclovododecaborane(12) from b9c2h11-2 through an insertion reaction. J Am Chem Soc 87(19):4392–4393 (b) Hawthorn MF Young DC, Andrews TD et al (1968) Pi-dicarbollyl derivatives of transition metals. Metallocene analogs. J Am Chem Soc 90(4):879–896 (c) Li J, Jones M (1990) A simple synthesis of 3-bromo-o-carborane. Inorg Chem 29(20):4162–4163 (d) Li J, Caparrelli DJ, Jones M (1993) 3-o-carboranylcarbenes. J Am Chem Soc 115(2):408–413 (e) Ramachandran BM, Knobler CB, Hawthorne MF (2006) Synthesis and structural characterization of symmetrical closo-4,7-i-2-1,2-$C_2B_{10}H_{10}$ and $[(CH_3)_3NH]$ [nido-2,4-i-2-7,8-$C_2B_9H_{10}$]. Inorg Chem 45(1):336–340
15. (a) Zi G, Li HW, Xie Z (2001) A novel carbons-adjacent arachno-C_2B_{10} carborane tetraanion bearing both hexagonal and pentagonal bonding faces. Organometallics 20(18):3836–3838 (b) Zi G, Li HW, Xie Z (2001) The first full-sandwich potassacarborane and a novel ‘carbons-adjacent’ $R_2C_2B_{10}H_{11}^-$ monoanion. Chem Comm 1110–1111 (c) Zi G, Li HW, Xie Z (2002) Synthesis, structural characterization, and reactivity of ‘carbons-adjacent’ nido- and arachno-carborane anions of the C_2B_{10} systems and their metal complexes. Organometallics 21(24):5415–5427 (d) Xie Z (2003) Controlled synthesis of carbons-adjacent and-apart nido-and arachno-carborane anions and their metal complexes. Pure Appl Chem 75(9):1335–1341
16. (a) Deng L, Xie Z (2007) A journey from 12-vertex to 14-vertex carboranes and to 15-vertex metallacarboranes. Organometallics 26(8):1832–1845 (b) Zhang J, Xie Z (2010) Recent progress in the chemistry of supercarboranes. Chem–Asian J 5(8):1742–1757

17. (a) Getman TD, Knobler CB, Hawthorne MF (1990) Multiple-bond character between two carborane polyhedral: the two-electron reduction of biscarborane. J Am Chem Soc 112 (11):4593–4594 (b) Fu X, Chan HS, Xie Z (2007) Synthesis and crystal structure of a 13-vertex carborane radical anion with 2n + 3 framework electrons. J Am Chem Soc 129 (29):8964–8965
18. (a) Dunks GB, Wiersema RJ, Hawthorn MF (1973) Chemical and structural studies of $B_{10}C_2H_{12}^-$ ions produced from icosahedral $B_{10}C_2H_{12}$ carboranes. J Am Chem Soc 95 (10):3174–3179 (b) Tolpin EI, Wong HS, Lipscomb WN (1974) Binding studies of boron hydride derivatives to proteins for neutron-capture therapy. J Med Chem 17(8):792–796 (c) Churchil.Mr, Deboer BG (1973) Crystal-structure and molecular geometry of [me4n +] [me2b10c2h11-], including refinement of positional and anisotropic thermal parameters for all hydrogen-atoms. Inorg Chem 12(11):2674–2682 (d) Chui K, Li HW, Xie Z (2000) Synthesis and structural characterization of closo-exo, exo-nido, and pseudocloso group 1 carborane compounds of the C_2B_{10} system. Organometallics 19(25):5447–5453
19. (a) Grimes RN (2000) Metallacarboranes in the new millennium. Coord Chem Rev 200:773–811 (b) Saxena AK, Hosmane NS (1993) Recent advances in the chemistry of carborane metal-complexes incorporating d-block and f-block elements. Chem Rev 93(3):1081–1124 (c) Saxena AK, Maguire JA, Hosmane NS (1997) Recent advances in the chemistry of heterocarborane complexes incorporating s- and p-block elements. Chem Rev 97(6):2421–2461 (d) Xie Z (2001) New class of metallacarboranes incorporating the eta(7)-carboranyl ligand. Pure Appl Chem 73(2):361–365 (e) Xie Z (2003) Cyclopentadienyl-carboranyl hybrid compounds: a new class of versatile ligands for organometallic chemistry. 36(1):1–9 (f) Hosmane NS, Maguire JA (2005) Evolution of C_2B_4 carborane chemistry: From early years to the present. Organometallics 24(7):1356–1389
20. Deng L, Xie Z (2007) A journey from 12-vertex to 14-vertex carboranes and to 15-vertex metallacarboranes. Organometallics 26(8):1832–1845
21. (a) Burke A, Ellis D, Giles BT et al (2003) Beyond the icosahedron: The first 13-vertex carborane. Angew Chem Int Ed 42(2):225–230 (b) Deng L, Chan HS, Xie Z (2005) Synthesis, reactivity, and structural characterization of a 14-vertex carborane. Angew Chem Int Ed 44 (14):2128–2131 (c) Deng L, Chan HS, Xie Z (2006) Synthesis, structure, and reactivity of 13-vertex carboranes and 14-vertex metallacarboranes. J Am Chem Soc 128(15):5219–5230
22. King RB (2001) Three-dimensional aromaticity in polyhedral boranes and related molecules. Chem Rev 101(5):1119–1152
23. Herzog A, Maderna A, Harakas GN et al (1999) A camouflaged nido-carborane anion: facile synthesis of octa-B-methyl-1,2-dicarba-closo-dodecaborane(12) and its deboration reaction. Chem-Eur J 5(4):1212–1217
24. Mironov VF, Grigos VI, Pechurin.Sy et al (1973) Production of a new class of silicon containing carboranes by interaction of vinyltrichlorosilane with ortho carboranes, meta carboranes para carboranes. Dokl Akad Nauk Sssr + 210(3):601–604
25. (a) Zakharki Li, Kalinin VN, Gedymin VV (1969) Synthesis and some reactions of 3-amino-o-carboranes. J Organomet Chem 16(3):371–379 (b) Zakharki Li, Kalinin VN, Gedymin VV et al (1970) Some reactions of 3-amino-ortho-carboranes. J Organomet Chem 23 (2):303–315
26. (a) Grushin VV, Shcherbina TM, Tolstaya TP (1985) The reactions of phenyl(B-carboranyl) iodonium salts with nucleophiles. J Organomet Chem 292(1–2):105–117 (b) Grushin VV, Demkina II, Tolstaya TP (1991) Reactions of aryl(B-carboranyl)iodonium cations with the fluoride anion: synthesis of icosahedral ortho-carboran-9-yl, meta-carboran-9-yl, and para-carboran-2-yl fluorides. Inorg Chem 30(25):4860–4863 (c) Grushin VV, Demkina II, Tolstaya TP (1992) Unified mechanistic analysis of polar reactions of diaryliodonium salts. J Chem Soc Perk T 2(4):505–511
27. Grushin VV (1992) Carboranylhalonium ions: from striking reactivity to a unified mechanistic analysis of polar reactions of diarylhalonium compounds. Acc Chem Res 25(11):529–536
28. Zakharkin LI, Kalinin VN, Kvasov BA et al (1968) Izv Akad Nauk SSSR Ser 2295–2298

29. (a) Bregadze VI, Kampel VT, Godovikov NN (1976) Mercuration of ortho-carboranes and meta-carboranes. J Organomet Chem 112(3):249–251 (b) Bregadze VI, Kampel VT, Godovikov NN (1977) B-mercurated carboranes. J Organomet Chem 136(3):281–287
30. Usyatinskii AY, Ryabov A D, Bregadze V I et al (1982) Izv Akad Nauk SSSR Ser 1598–1603
31. (a) Hoel EL, Hawthorn MF (1973) Intramolecular oxidative addition of iridium to a boron-hydrogen bond of a complexed carboranylphosphine. J Am Chem Soc 95(8):2712–2713 (b) Hoel EL, Hawthorn MF (1974) Preparation of beta-sigma-metallocarboranes by oxidative addition—model intermediates for transition-metal catalyzed hydrogen-deuterium exchange in boron-hydrogen containing species. J Am Chem Soc 96(21):6770–6771 (c) Hoel EL, Hawthorne MF (1975) Preparation of b-sigma-carboranyl iridium complexes by oxidative addition of terminal boron-hydrogen bonds to iridium(i) species. J Am Chem Soc 97(22):6388–6395
32. (a) Peymann T, Herzog A, Knobler CB et al (1999) Aromatic polyhedral hydroxyborates: Bridging boron oxides and boron hydrides. Angew Chem Int Ed 38(8):1062–1064 (b) Herzog A, Knobler CB, Hawthorne MF (2001) Synthesis, structure, and reactivity of closo-2,3,4,5,6,7,8,9,10,11-decahydroxy-1,12-bis(sulfonic acid)-1,12-dicarbadodecaborane(12). J Am Chem Soc 123(51):12791–12797
33. Zakharkin LI, Kalinin VN, Podvisotskaya LS (1965) Izv. Akad. Nauk. SSSR, Ser. Khim. 1684–1686 (b) Zakharkin LI, Kalinin VN, Podvisotskaya LS (1966) Zh. Obshch. Khim. 36:1779–1781
34. Zheng G, Jones M (1983) Reaction of carbethoxycarbene with o-carborane. J Am Chem Soc 105(21):6487–6488
35. (a) Sung DD, Lee JD, Choi SK (1987) A study on the reaction of icosahedral carborane with carbenes. B Kor Chem Soc 8(2):63–68 (b) Yuan KX, Jones M (1992) Carbenes do react with p-carborane. Tetrahedron Lett 33(49):7481–7484
36. (a) Wu S, Jones M (1986) Carbon-to-boron bridged ortho-carboranes. Inorg Chem 25 (27):4802–4803 (b) Wu S, Jones M (1988) A carbon-to-boron-bridged benzo-o-carborane. Inorg Chem 27(11):2005–2008 (c) Wu S, Jones M (1989) The mechanism of rearrangement of the icosahedral carboranes. J Am Chem Soc 111(14):5373–5384
37. Hermansson K, Wojcik M, Sjoberg S (1999) O-, m-, and p-carboranes and their anions: ab initio calculations of structures, electron affinities, and acidities. Inorg Chem 38(26):6039–6048
38. Bregadze VI (1992) Dicarba-closo-dodecaboranes $C_2B_{10}H_{12}$ and their derivatives. Chem Rev 92(2):209–223
39. Nakamura H, Aoyagi K, Yamamoto Y (1998) Tetrabutylammonium fluoride promoted novel reactions of o-carborane: inter- and intramolecular additions to aldehydes and ketones and annulation via enals and enones. J Am Chem Soc 120(6):1167–1171
40. Nakamura H, Yamamoto Y (1999) Novel addition and [3 + 2] cycloaddition reactions of stannyl- and silyl-ortho-carboranes to carbonyl compounds. Collect Czech Chem C 64 (5):829–846
41. (a) Bregadze VI, Usyatinskii AY, Kampel VT et al (1985) Carboranyl derivatives of group-iii non-transition metals. Bulletin of the Academy of Sciences of the Ussr Division of Chemical Science 34(5):1113 (b) Bregadze VI, Kampel VT, Usyatinskii AY et al (1978) Thalloborated carboranes. Bul Acad Sci Ussr Div Chem Sci 27(6):1284
42. (a) Lee JD, Baek CK, Ko JJ et al (1999) Syntheses and crystal structures of intramolecularly stabilized organogallium compounds containing an o-carboranyl C,N-chelating ligand system. Organometallics 18(11):2189–2197 (b) Lee JD, Ko J, Cheong M et al (2005) Syntheses and crystal structures of intramolecularly stabilized organo aluminum, gallium, and indium compounds containing the C,P-chelating o-carboranylphosphino ligand [o-$C_2B_{10}H_{10}$ (CH_2PMe_2)-C,P]$^-$ ($Cab^{C,P}$). X-ray structure of pentacoordinated group 13 metal complexes $(Cab^{C,P})_2MX$ (M = Ga, in; X = Cl). Organometallics 24(24):5845-5852
43. Fedorov LA, Kalinin VN, Gasanov KG et al (1975) Zh Obshch Khim 45:591–607
44. Bregadze VI (1992) Dicarba-closo-dodecaboranes $C_2B_{10}H_{12}$ and their derivatives. Chem Rev 92(2):209–223

45. (a) Gomez FA, Johnson SE, Hawthorne MF (1991) A versatile protecting group for 1,2-dicarba-closo-dodecaborane(12) and the structure of a silylcarborane derivative. J Am Chem Soc 113(15):5915–5917 (b) Gomez FA, Hawthorne MF (1992) A simple route to c-monosubstituted carborane derivatives. J Org Chem 57(5):1384–1390
46. Zi G, Li HW, Xie Z (2002) Synthesis, structural characterization, and reactivity of organolanthanide complexes derived from a new, versatile boron-bridged ligand, $^{i}Pr_2NB(C_9H_7)(C_2B_{10}H_{11})$. Organometallics 21(6):1136–1145
47. Brown DA, Colquhoun HM, Daniels JA et al (1992) Polymers and ceramics based on icosahedral carboranes: model studies of the formation and hydrolytic stability of aryl ether, ketone, amide and borane linkages between carborane units. J Mater Chem 2(8):793–804
48. (a) Heying TL, Ager JW, Clark SL et al (1963) A new series of organoboranes.III. Some reactions of 1,2-dicarbaclovododecaborane(12) and its derivatives. Inorg Chem 2(6):1097–1105 (b) de Rege FM, Kassebaum JD, Scott BL et al (1999) Synthesis, structure, and reactivity of 1,2-(1′,1′,2′,2′-tetramethyldisilane-1,2′)carborane. Inorg Chem 38(3):486–489 (c) Song KH, Jung I, Lee SS et al (2001) Palladium-catalyzed double-silylation reactions of 3,4-carboranylene-1,1,1,2-tetraethyl-1,2-disilacyclobut-3-ene. Organometallics 20(26):5537–5541
49. (a) Alexander RP, Schroeder H (1963) Chemistry of decaborane-phosphorus compounds. IV. Monomeric, oligomeric, and cyclic phosphinocarboranes. Inorg Chem 2(6):1107–1110 (b) Zakharkin LI, Brattsev VA, Chapovsky YA et al (1964) Some special features of structure and reactivity of barene compounds. Dokl Akad Nauk Sssr 157(5):1149–1151 (c) Kazantse.Av, Zhubekov.Mn, Zakharki.Li (1971) Synthesis and conversions of substituted ortho carboranylphenylchlorophosphines and bis(ortho carboranyl)chlorophosphines. Zh Obshch Khim 41(9):2027–2033 (d) Bregadze VI, Godovikov NN, Degtyarev AN et al (1976) Synthesis of tris(methyl-ortho-carboranyl)phosphine. J Organomet Chem 112(2): C25–C26
50. Heying TL, Ager JW, Clark SL et al (1963) A new series of organoboranes. 3. Some reactions of 1,2-dicarbaclovododecaborane(12) and its derivatives. Inorg Chem 2(6):1097–1105
51. (a) Stanko VI, Anorova GA (1967) Reaction of o-barenylmagnesium bromide with carbon tetrachloride. J Gen Chem Ussr 37(2):476–477 (b) Stanko VI, Anorova GA (1968) Preparation of 1-chloro-o-barene. J Gen Chem Ussr 38(6):1356–1360 (c) Stanko VI, Anorova GA (1970) Chlorination of ortho barenylmagnesium bromides: determination of C,B-dichloro ortho barene structure. Zh Obshch Khim 40(3):646–647
52. (a) Gingrich HL, Ghosh T, Huang Q et al (1990) 1,2-dehydro-ortho-carborane. J Am Chem Soc 112(10):4082–4083 (b) Jeon J, Kitamura T, Yoo BW et al (2001) Synthesis and reactivity of an efficient 1,2-dehydrocarborane precursor, phenyl[o-(trimethylsilyl)carboranyl]iodonium acetate. Chem Comm 20:2110–2111
53. (a) Qiu Z, Wang SR, Xie Z (2010) Nickel-catalyzed regioselective [2 + 2 + 2] cycloaddition of carboryne with alkynes. Angew Chem Int Edit 49(27):4649–4652 (b) Qiu Z, Xie Z (2014) Generation and reactivity of o-carborynes. Dalton Trans 43(13):4925–4934
54. Wang SR, Qiu Z, Xie Z (2011) Regioselective insertion of carborynes into ethereal C-H bond: facile synthesis of alpha-carboranylated ethers. J Am Chem Soc 133(15):5760–5763
55. (a) Ghosh T, Gingrich HL, Kam CK et al (1991) Reactions of 1,2-dehydro-ortho-carborane with dienes. J Am Chem Soc 113(4):1313–1318 (b) Gingrich HL, Huang Q, Morales AL et al (1992) Reactions of enamines with dehydro aromatic compounds. J Org Chem 57(14):3803–3806 (c) Huang QR, Gingrich HL, Jones M (1991) Ene reactions of 1,2-dehydro-ortho-carborane. Inorg Chem 30(17):3254–3257
56. Hoffmann RW (1967) Dehydrobenzene and cycloalkynes. Academic Press, New York
57. Lee T, Jeon J, Song KH et al (2004) Generation and trapping reaction of an efficient 1,2-dehydrocarborane precursor, phenyl[o-(trimethylsilyl)carboranyl]iodonium acetate. Dalton Trans 6:933–937
58. (a) Tadross PM, Stoltz BM (2012) A comprehensive history of arynes in natural product total synthesis. Chem Rev 112(6):3550–3577 (b) Pena D, Perez D, Guitian E (2006) Insertion of

arynes into sigma bonds. Angew Chem Int Ed 45(22):3579–3581 (c) Wenk HH, Winkler M, Sander W (2003) One century of aryne chemistry. Angew Chem Int Ed 42(5):502–528
59. Atkins JH, Ho DM, Jones M (1996) A new source of 1,2-dehydro-o-carborane. Tetrahedron Lett 37(40):7217–7220
60. Wang SR, Xie Z (2012) Reaction of carboryne with alkylbenzenes. Organometallics 31 (8):3316–3323
61. Wittig G (1959) Triptycene 39:75–77
62. Brinkley JM, Friedman L (1972) Novel ene and insertion reactions of benzyne with alkylbenzenes. Tetrahedron Lett 40:4141–4142
63. BarnettThamattoor L, Zheng GX, Ho DM et al (1996) Reactions of 1,2-dehydro-o-carborane with thiophenes. Cycloadditions and an easy synthesis of "benzo-o-carboranes". Inorg Chem 35(25):7311–7315
64. Wang SR, Xie Z (2012) Reaction of carboryne with styrene and its derivatives. Tetrahedron 68 (26):5269–5278
65. Ciganek E (1969) Negatively substituted acetylenes. II. Cycloaddition reactions with styrenes. J Org Chem 34(6):1923–1930
66. Ahlgren G, Akerman B (1970) The addition of benzyne to 1, 2-dideuteriocyclohexene. Tetrahedron Lett 11:3047–3048
67. Kuehn ME (1962) The arylation of enamines. J Am Chem Soc 84(5):837–847
68. (a) Cunningham RJ, Bian NY, Jones M (1994) Reaction of 1,2-dehydro-o-carborane with acetylenes - ene reactions and cycloaddition to give an analog of benzocyclobutadiene. Inorg Chem 33(22):4811–4812 (b) Ho DM, Cunningham RJ, Brewer JA et al (1995) Reaction of 1,2-dehydro-o-carborane with acetylenes—synthesis and structure of a carborane analog of benzocyclobutadiene. Inorg Chem 34(21):5274–5278
69. Wang SR, Qiu Z, Xie Z (2010) Insertion of carboryne into aromatic rings: formation of cyclooctatetraenocarboranes. J Am Chem Soc 132(29):9988–9989
70. Bender CO, Cassis IM, Dolman D et al (1984) Evidence for a di-pi-methane rearrangement in the photoisomerization of 2,3-benzo-7,8-dimethylbicyclo[4.2.0]octa-2,4,7-triene to 6,7-dimethylbenzotricyclo[3.3.0.02,8]octa-3,6-diene. Can J Chem 62(12):2769–2776
71. Wang SR, Qiu Z, Xie Z (2011) Regioselective insertion of carborynes into ethereal C-H bond: facile synthesis of alpha-carboranylated ethers. J Am Chem Soc 133(15):5760–5763
72. (a) Okuma K, Fukuzaki Y, Nojima A et al (2008) Novel tandem reaction of benzyne with cyclic ethers and active methines: Synthesis of omega-trichloroalkyl phenyl ethers. Tetrahedron Lett 49(19):3063–3066 (b) Feltenberger JB, Hayashi R, Tang Y et al (2009) Enamide-benzyne-[2 + 2] cycloaddition: Stereoselective tandem [2 + 2]-pericyclic ring-opening-intramolecular n-tethered [4 + 2] cycloadditions. Org Lett 11(16):3666–3669
73. (a) Pan SG, Liu JH, Li HR et al (2010) Iron-catalyzed n-alkylation of azoles via oxidation of C-H bond adjacent to an oxygen atom. Org Lett 12(9):1932–1935 (b) Tzirakis MD, Orfanopoulos M (2010) Photochemical addition of ethers to C60: Synthesis of the simplest [60] fullerene/crown ether conjugates Angew Chem Int Ed 49(34):5891–5893
74. Woodward RB, Rosenblum M, Whiting MC (1952) A new aromatic system. J Am Chem Soc 74(13):3458–3459
75. Wang SR, Xie Z (2012) Formal insertion of o-carborynes into ferrocenyl C-H bonds: a simple access to o-carboranylferrocenes. Organometallics 31(12):4544–4550
76. Qiu Z, Ren S, Xie Z (2011) Transition metal-carboryne complexes: synthesis, bonding, and reactivity. Acc Chem Res 44(4):299–309
77. (a) Sayler AA, Sieckhau JF, Beall H (1973) Unusual chelated o-carborane transition-metal complex. J Am Chem Soc 95(17):5790–5792 (b) Qiu Z, Deng L, Chan HS et al (2010) Synthesis and structural characterization of group 10 metal-carboryne complexes. Organometallics 29(20):4541–4547
78. (a) Wang H, Li HW, Huang X et al (2003) Synthesis, structure, and bonding of a zirconocene-1,2-dehydro-o-carborane complex. Angew Chem Int Ed 42(36):4347–4349 (b) Ren S, Deng L, Chan HS et al (2009) Synthesis and structural characterization of zirconium-carboryne complexes. Organometallics 28(19):5749–5756

79. (a) Deng L, Chan HS, Xie Z (2006) Nickel-mediated regioselective [2 + 2 + 2] cycloaddition of carboryne with alkynes. J Am Chem Soc 128(24):7728–7729 (b) Qiu Z, Xie Z (2008) Nickel-mediated coupling reactions of carboryne with alkenes: A synthetic route to alkenylcarboranes. Angew Chem Int Edit 47(35):6572–6575 (c) Qiu Z, Xie Z (2009) Nickel-mediated three-component cycloaddition reaction of carboryne, alkenes, and alkynes. J Am Chem Soc 131(6):2084–2085
80. Qiu Z, Wang SR, Xie Z (2010) Nickel-catalyzed regioselective [2 + 2 + 2] cycloaddition of carboryne with alkynes. Angew Chem Int Edit 49(27):4649–4652
81. (a) Deng L, Chan HS, Xie ZW (2005) Synthesis, structure, and reactivity of a zirconocene-carboryne precursor. J Am Chem Soc 127(40):13774–13775 (b) Ren S, Chan HS, Xie Z (2009) Synthesis, structure, and reactivity of zirconacyclopentene incorporating a carboranyl unit. J Am Chem Soc 131(11):3862–2863
82. Ren S (2010) Synthesis, structure, and reactivity of group 4 metal-carboryne complexes. Ph.D. Thesis, The Chinese University of Hong Kong, Hong Kong
83. (a) Jemmis ED, Kiran B (1997) Structure and bonding in $B_{10}X_2H_{10}$ (x = C and Si). The kinky surface of 1,2-dehydro-o-disilaborane. J Am Chem Soc 119(17):4076–4077 (b) Kiran B, Anoop A, Jemmis ED (2002) Control of stability through overlap matching: Closo-carborynes and closo-silaborynes. J Am Chem Soc 124(16):4402–4407
84. Qiu Z, Xie Z (2010) Palladium/nickel-cocatalyzed cycloaddition of 1,3-dehydro-o-carborane with alkynes. Facile synthesis of c, b-substituted carboranes. J Am Chem Soc 132(45): 16085–16093

Chapter 2
Regioselective Insertion of *o*-Carborynes into α-C–H Bond of Tertiary Amines

Abstract This chapter describes the regioselective insertion of *o*-carborynes into the *α*-C–H bond of amines. Different from our previous work on the reaction of *o*-carborynes with aliphatic ethers, the bonding form of *o*-carboryne is involved in this reaction. Nucleophilic addition of tertiary amines to the multiple bond of *o*-carboryne generates a zwitterionic intermediate, which undergoes a sequence of intramolecular proton transfer and nucleophilic attack to afford the final insertion product. This serves as a simple and general methodology for the generation of a series of 1-aminoalkyl-*o*-carboranes.

2.1 Introduction

α-Functionalization of tertiary amines has received considerable attention due to the increasing interest in carbon–carbon bond formation [1] and the importance of amines and alkaloids in medicinal chemistry [2]. To achieve this goal, different strategies such as lithiation method [3], amine radical formation [4], and transition-metal-catalyzed C–H bond activation [5] have been developed. Among them, iminium ions are one of the key intermediates for the construction of amine stereocenters [5d, 6].

On the other hand, a growing interest has been directed toward the construction of *o*-carborane derivatives bearing nitrogenated substituents due to their potential application in medicinal chemistry [7] and catalysis [8]. For instance, carborane–amino acid or carborane–nucleoside combinations serve as excellent candidates for cancer treatment in boron neutron capture therapy (BNCT) [7b–e, 9]. Moreover, aminoalkyl-*o*-carboranes have been extensively employed as ligands for the synthesis and characterization of metal complexes [8, 10]. Despite the remarkable progress in carborane chemistry, straightforward and general synthesis of such aminoalkyl-*o*-carboranes still represents a very challenging task [11]. In this section, a highly regioselective α-carboranylation of tertiary amines for the convenient synthesis of carborane-containing amines will be discussed.

D. Zhao, *Functionalization of Carborane via Carboryne Intermediates*,
Springer Theses, DOI 10.1007/978-981-10-1569-4_2

2.2 Reaction of *o*-Carborynes with Tertiary Amines

We have reported that *o*-carboryne (**II-3a**) [12], a very reactive intermediate, can insert into the aliphatic ethereal α-C–H bond with the selectivity of secondary C–H >> primary C–H (Scheme 2.1) [13]. In view of that the bond dissociation energy of amine α-C–H bond is relatively smaller than that of ether α-C–H bond [14], we wondered whether α-functionalization of amine could be achieved in a similar way.

The reaction of 1-I-2-Li-*o*-carborane (**II-2a**) with triethylamine (**II-4a**) was chosen as a model to establish the reaction conditions (Table 2.1). Heating an *n*-hexane suspension of **II-2a** in the presence of 20 equiv of **4a** at 60 °C for 1 h afforded only the α-carboranylated amine **II-5aa** in 67 % isolated yield after workup (Table 2.1, entry 1). Lowering the amount of **II-4a** to 10 equiv gave **II-5aa** in a better yield of 72 % (Table 2.1, entries 2–3). Carrying out the reaction temperature at 80 °C did not improve the yield (Table 2.1, entry 4). Heating is necessary for this reaction as only trace amount of **II-5aa** was formed at room temperature for 12 h (Table 2.1, entry 6). These results suggest that *o*-carboryne favors insertion into the α-C–H bond of amine with the formation of the C–C bond rather than the C–N bond that is observed in the reaction of benzyne (Scheme 2.1) [15].

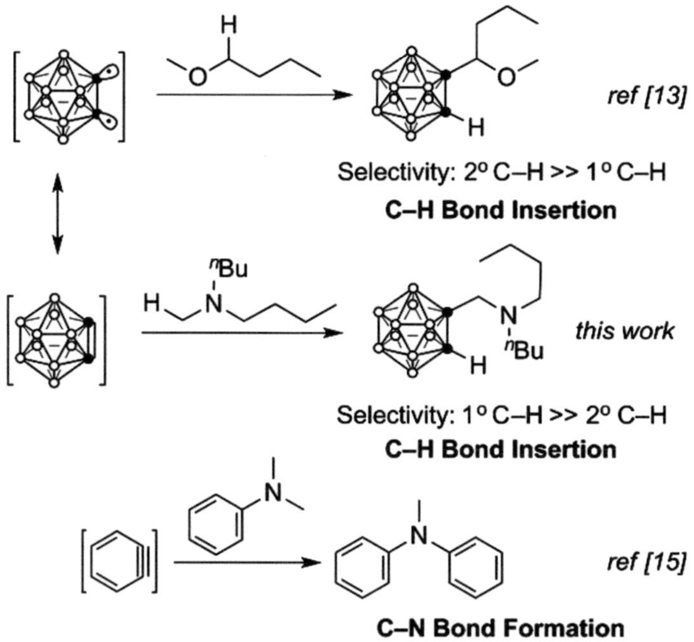

Scheme 2.1 Reaction of **II-3a** with ethers and tertiary amines

Table 2.1 Optimization of reaction conditions[a]

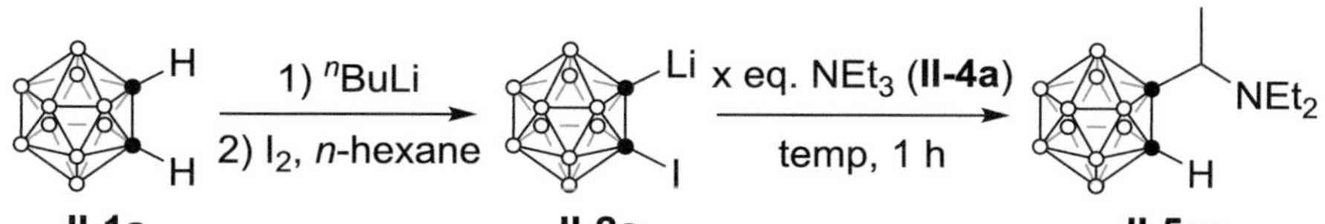

Entry	**4a** (equiv)	Temp (°C)	**II-5aa** (%)[b]
1	20	60	67
2	10	60	72
3	5	60	55[c]
4	10	80	70
5	10	40	58[c]
6[d]	10	rt	Trace

[a]Reaction conditions: **II-1a** (1.0 mmol), ^{n}BuLi (2.0 mmol), I_2 (1.0 mmol), *n*-hexane (10 mL), rt, 15 min, then **II-4a** (5.0, 10.0, or 20.0 mmol)
[b]Isolated yield
[c]**II-2a** was not completely consumed
[d]12 h

Under the optimal reaction conditions depicted in Table 2.1, entry 2, the substrate scope was examined and the results were compiled in Table 2.2. Similar to triethylamine **II-4a**, symmetrical amines **II-4b–d** worked well with decreased yields, which might be associated with the increased steric hindrance (Table 2.2, entries 1–4). The structures of **II-5b** and **II-5d** were further confirmed by single-crystal X-ray analyses (Figs. 2.1 and 2.2). Although both GC/MS and NMR spectra indicated a very high conversion, the reaction of triallylamine **II-4e** resulted in only 45 % isolated yield probably due to the loss in purification on silica gel (Table 2.2, entry 5). *N*,*N*-Diisopropylethylamine **II-4f** was a poor substrate because of large steric hindrance caused by the two isopropyl groups (Table 2.2, entry 6). For asymmetrical amines, intriguing regioselective preferences were observed. For example, *N*,*N*-dimethylbutylamine **II-4g** or *N*-methyldibutylamine **II-4h** underwent α-carboranylation in good yields with a >14:1 or 1.8:1 selectivity for methyl C–H bond cleavage over secondary C–H bond cleavage, presumably reflecting an inherent ~>4.7:1 or ~2.4:1 selectivity, respectively (Table 2.2, entries 7–8). An exclusive preference for α-carboranylation at benzylic position over aliphatic position was observed for benzylamines **II-4i–l** (Table 2.2, entries 9–12). A fluoro substitution at the para position of the benzyl group increased the yield, while a methoxy substituent rendered the reaction less efficient (Table 2.2, entries 11–12). The above results indicate that α-carboranylation occurs preferentially at acidic positions or more electron-deficient C–H bonds (methyl vs. ^{n}butyl; benzyl vs. aliphatic). For cyclic amines, the reaction was much dependent on the ring size due probably to steric reasons [16]. For example, 5-membered cyclic amine **II-4m** was found to undergo α-carboranylation with a ~3.7:1 selectivity for cyclic positions

Table 2.2 α-Carboranylation of amines[a]

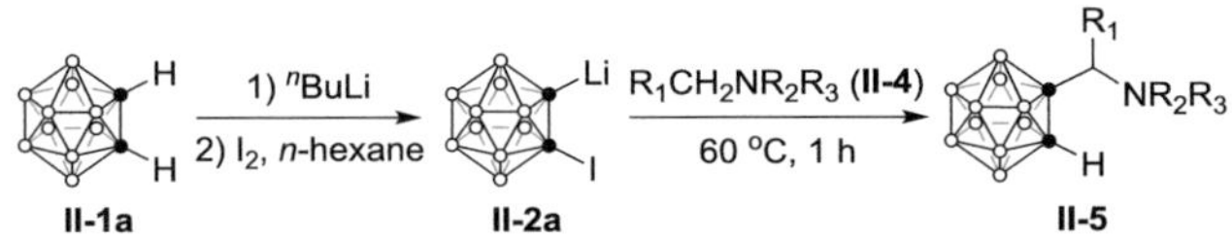

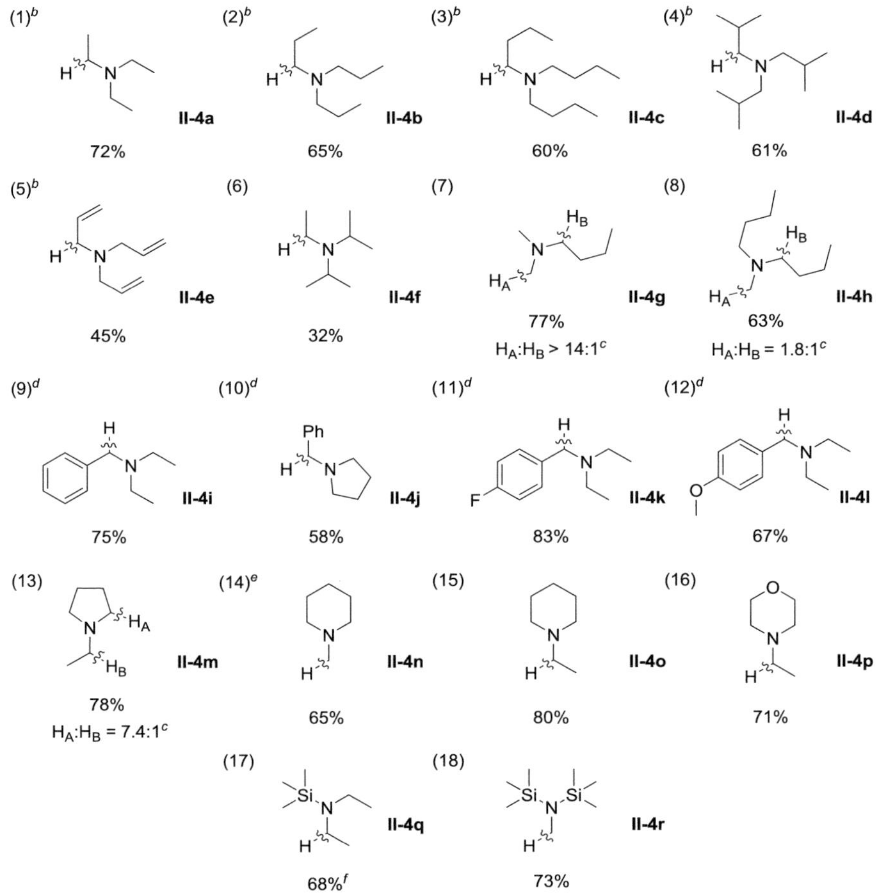

[a]Reaction conditions: **II-1a** (1.0 mmol), ^{n}BuLi (2.0 mmol), I_2 (1.0 mmol), *n*-hexane (10 mL), rt, 15 min, then **II-4** (10.0 mmol). Isolated yield was given

[b]1 h

[c]Determined by ^{1}H NMR

[d]2.5 equiv of amine was used

[e]**II-2a** was not completely consumed

[f]Desilylation occurred during isolation

over aliphatic chain positions (Table 2.2, entry 13), whereas the cyclic positions of 6-membered cyclic amines **II-4n** and **II-4o** were completely inactive (Table 2.2, entries 14–15). Similarly, the reaction of *N*-ethylmorpholine **II-4p** furnished only ethyl carboranylated amine (Table 2.2, entry 16), and no ethereal C–H cleavage

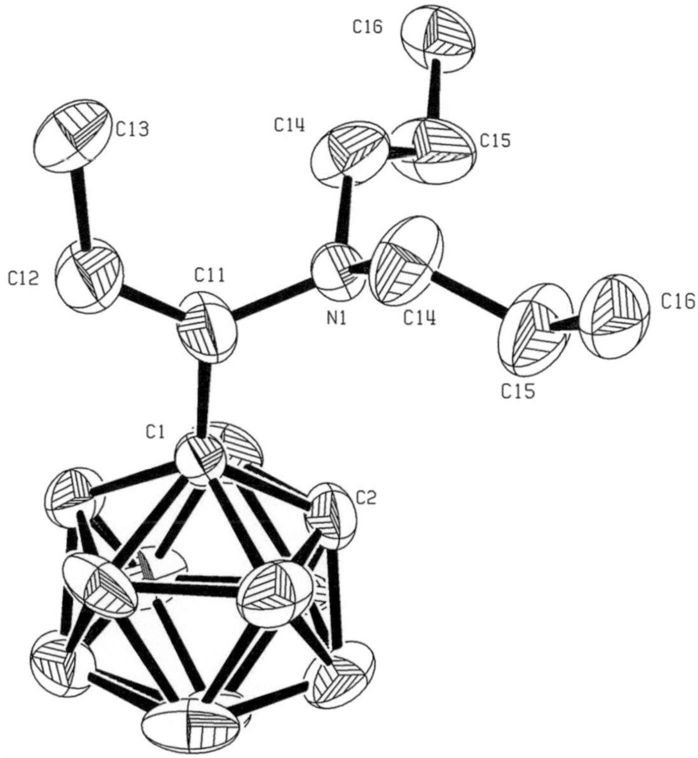

Fig. 2.1 Molecular structure of **II-5b**

was observed. *N*,*N*-diethyltrimethylsilylamine **II-4q** or *N*,*N*-bistrimethylsilylmethyl amine **II-4r** also worked well to afford **II-5q** or **II-5r** in % or 73 % isolated yield, respectively (Table 2.2, entries 17–18). The molecular structures of **II-5b**, **II-5d**, and **II-5f** were further confirmed by single-crystal X-ray analyses (Figs. 2.1, 2.2, and 2.3).

Interestingly, in the reaction of **II-4s**, a 1,2-disubstituted carborane **II-4s** was isolated as sole product (Scheme 2.2) and it was further confirmed by single-crystal X-ray analyses (Fig. 2.4). When aryl amine, such as **II-4t**, was examined in this reaction, the desired sp^3 C–H insertion product was not detected; instead, a mixture of *ortho* and *para* sp^2 C–H insertion products was formed in very good yields (Scheme 2.2). Such different reactivity of aliphatic amine and aryl amine toward *o*-carboryne may be ascribed to the different nucleophilicity of the amine unit.

Effects of cage B-substituents on α-carboranylation were also examined (Table 2.3). Two diastereoisomers were formed in a 1:1 ratio for 3-substituted

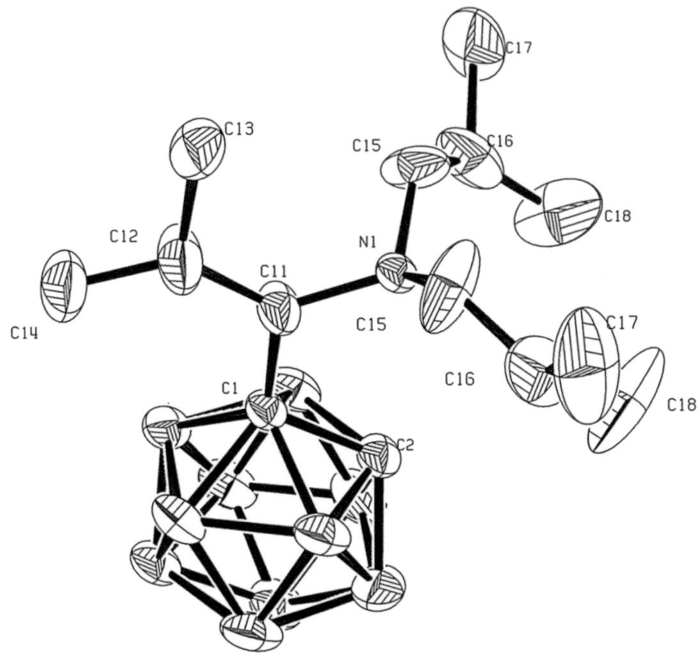

Fig. 2.2 Molecular structure of **II-5d**

carboranes (Table 2.3, entries 1–2). A dramatic influence of electronic effects was observed (Table 2.3, entries 3–5). For example, 9,12-dimethyl carboranylated amine **II-5da** was obtained in 76 % isolated yield, whereas no reaction was observed for 9,12-diiodo-*o*-carborane **II-1e** and 4,5,7,8,9,10,11,12-octamethyl-*o*-carborane **II-1f**.

2.3 Mechanistic Studies on *α*-Carboranylation of Amines

To gain some insight into the reaction pathway, several control experiments were performed (Scheme 2.3). The reaction proceeded well either under UV irradiation (365 nm) or in the dark (Eq. 2.1). The desired α-carboranylated amine **II-5aa** was still isolated in 62 % yield in the presence of 1.1 equiv of a radical scavenger, 2,2,6,6-tetramethylpiperidine-1-oxyl (TEMPO), indicating that a radical pathway was not necessarily involved (Eq. 2.2). In addition, no desired product was detected from the reaction of 1-I-7-Li-*m*-carborane (**II-m-2a**) with **II-4a** (Eq. 2.3). These experimental results, together with the regioselectivity observed in Table 2.2, are significantly different from those reported in our previous work [13] on the reaction of *o*-carboryne with ethers, which probably suggests the absence of the biradical form of *o*-carboryne in these reactions.

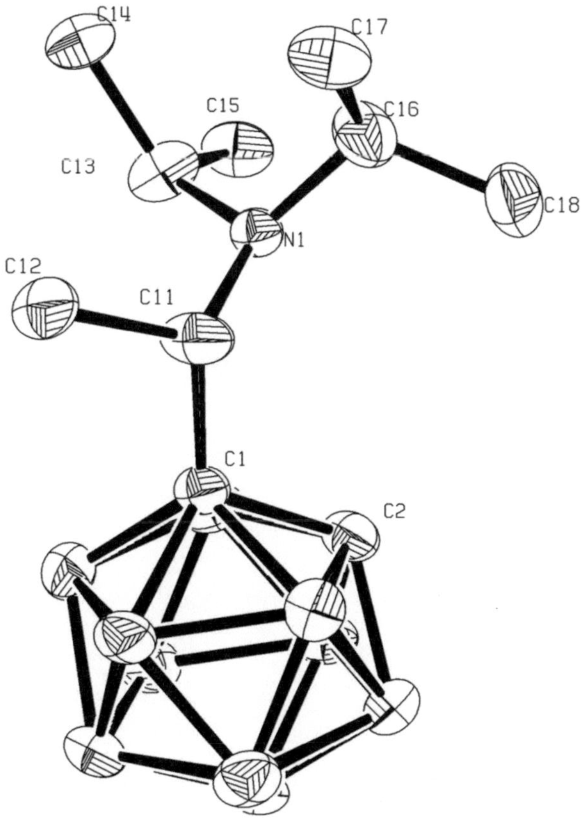

Fig. 2.3 Molecular structure of **II-5f**

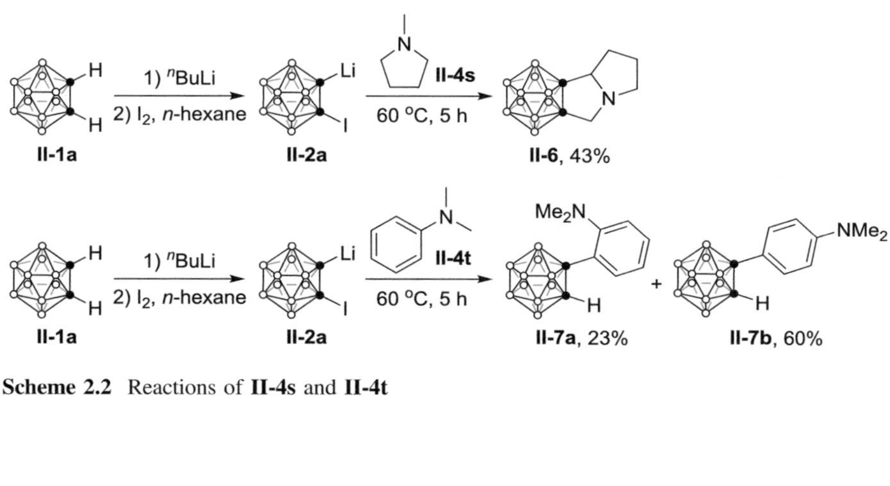

Scheme 2.2 Reactions of **II-4s** and **II-4t**

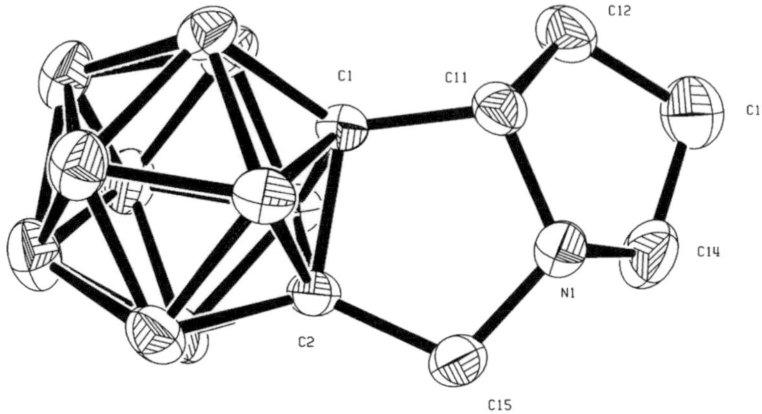

Fig. 2.4 Molecular structure of **II-6**

Table 2.3 Effects of cage B-substituents on α-carboranylation[a]

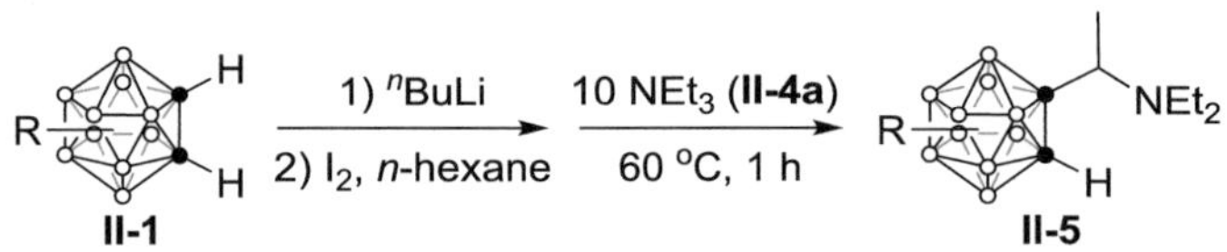

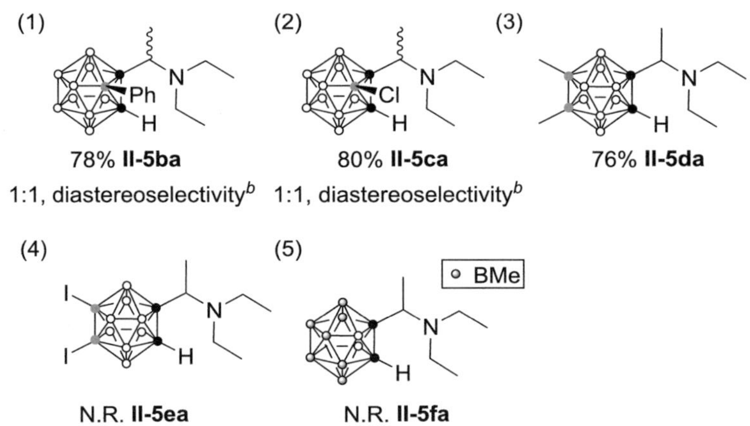

[a]Reaction conditions: **II-1** (1.0 mmol), ^{n}BuLi (2.0 mmol), I_2 (1.0 mmol), *n*-hexane (10 mL), rt, 15 min, then **II-4a** (10.0 mmol). Isolated yield was given

[b]Determined by ^{1}H NMR

Scheme 2.3 Control experiments

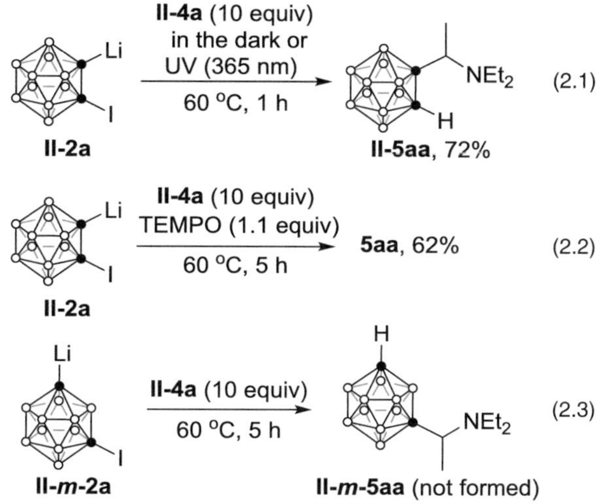

To determine the proton source, a deuterium-labeled substrate **II-4j-D** (>99 % D) was subjected to the reaction. Product **II-5j-D** with 87 % deuterium incorporation was obtained in 53 % yield (Scheme 2.4). This finding reveals that D (H) incorporated into the cage H mainly comes from the amine substrate (Figs. 2.5, 2.6, and 2.7).

Meanwhile, the intermolecular primary kinetic isotope effect (KIE) experiments were also carried out under standard conditions (Scheme 2.4). The results showed a k_H/k_D value of 1.04 (Figs. 2.8 and 2.9), which revealed that the C–H bond cleavage was not the rate-determining step.

2.4 DFT Calculations on α-Carboranylation of Amines

To gain further insight into the reaction pathway and to understand the origins of the selectivity for α-carboranylation of tertiary amines, DFT calculations were carried out (Fig. 2.10). All of these calculations were carried out with the Gaussian 09 program [17]. Geometry optimizations were performed at the Becke3LYP (B3LYP) [18] level of the density functional theory. The LANL2DZdp [19] basis set was used for I, while 6-31 + G(d,p) basis set was used for all other atoms. Frequency calculations were made to determine the characteristics of all stationary points as energy minima or transition states and obtain thermal corrections. Intrinsic reaction coordinates (IRCs) [20] were calculated to confirm that transition states lead to relevant intermediates. The solvent effect of *n*-hexane was taken into consideration by performing single-point solvation energy calculations with the conductor-like polarizable continuum model (CPCM) [21] using UAKS radii.

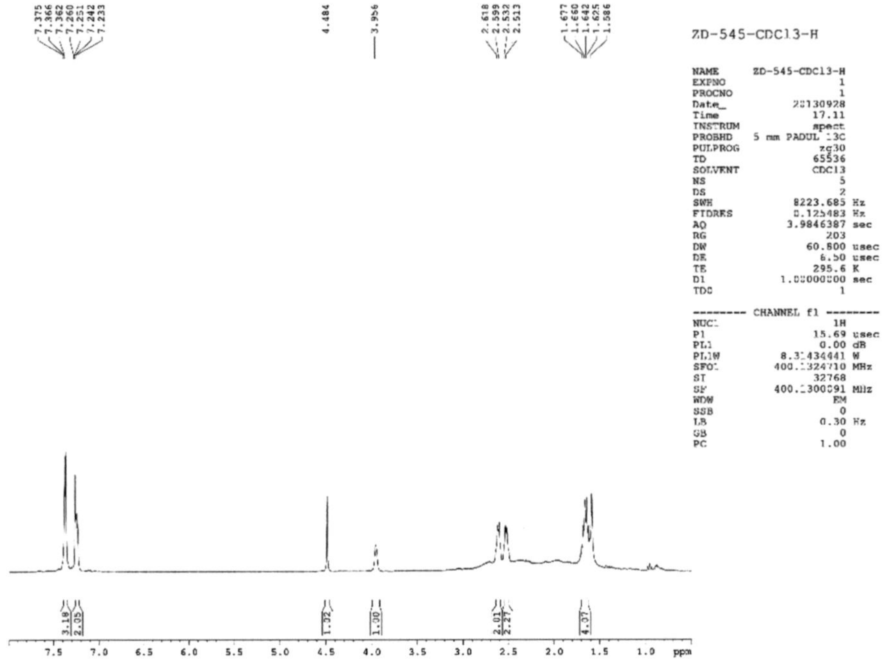

Fig. 2.5 ^{1}H NMR spectrum of **II-5j** (for reference)

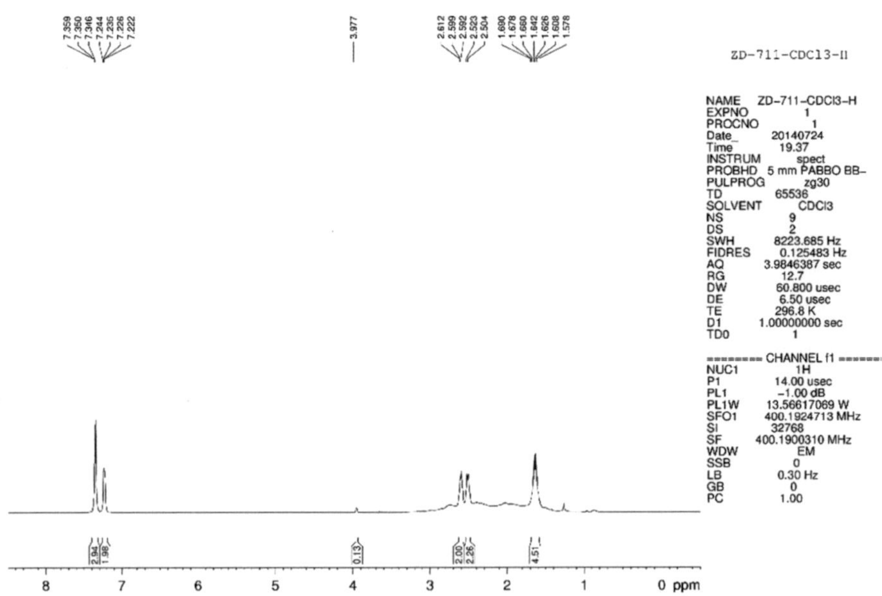

Fig. 2.6 ^{1}H NMR spectrum of **II-5j-D$_2$**

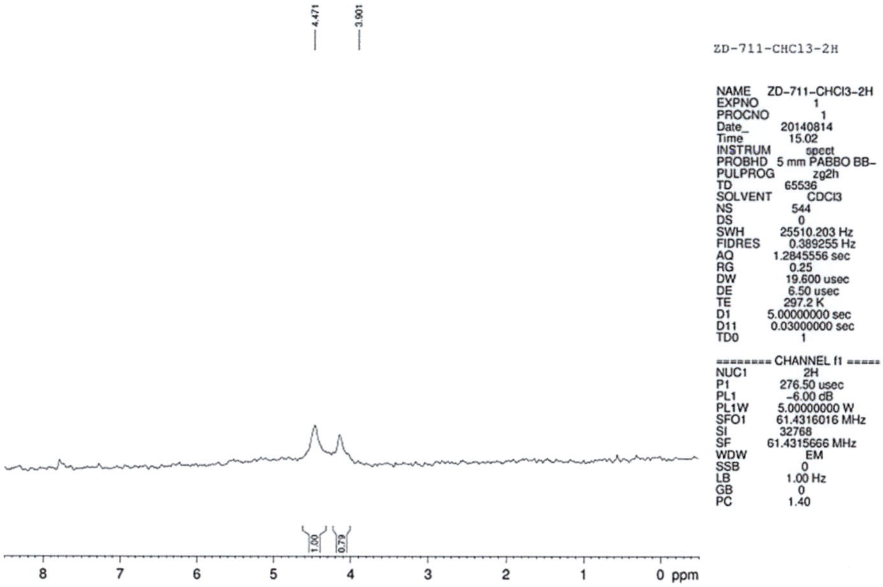

Fig. 2.7 ^{2}H NMR spectrum of **II-5j-D$_2$**

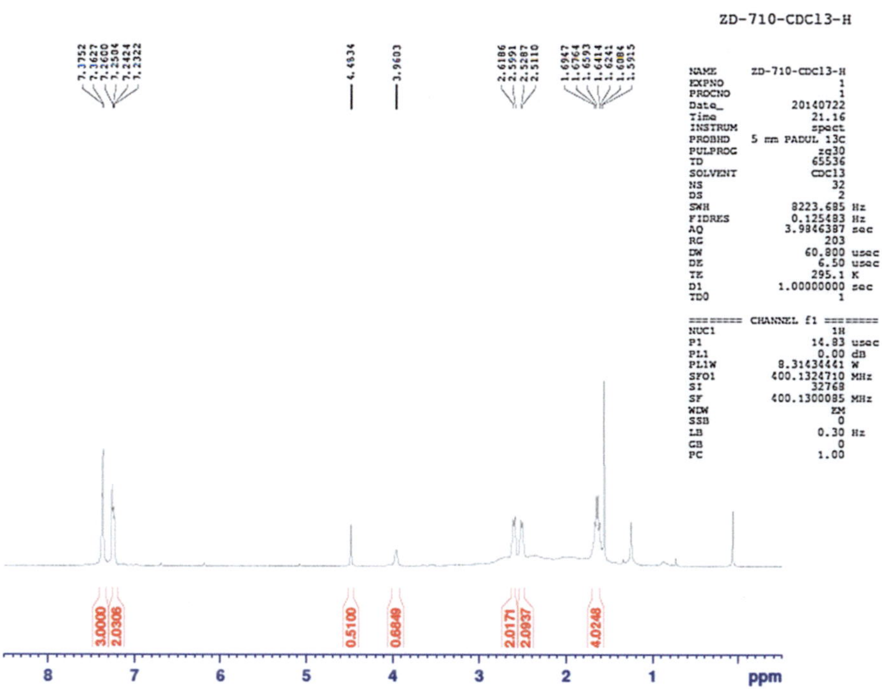

Fig. 2.8 ^{1}H NMR spectrum for determination of *KIE* value

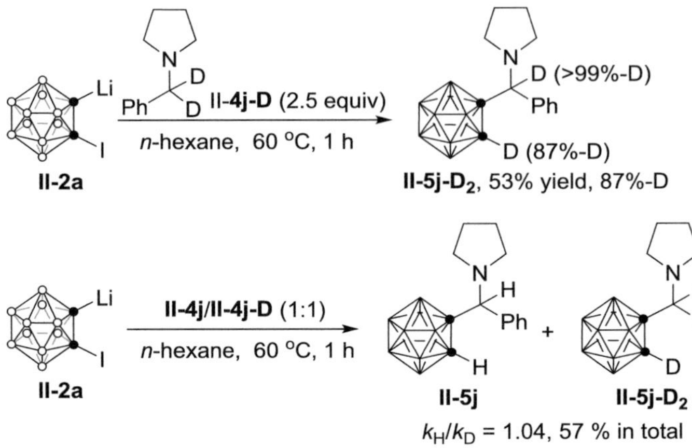

Scheme 2.4 Deuterium experiments

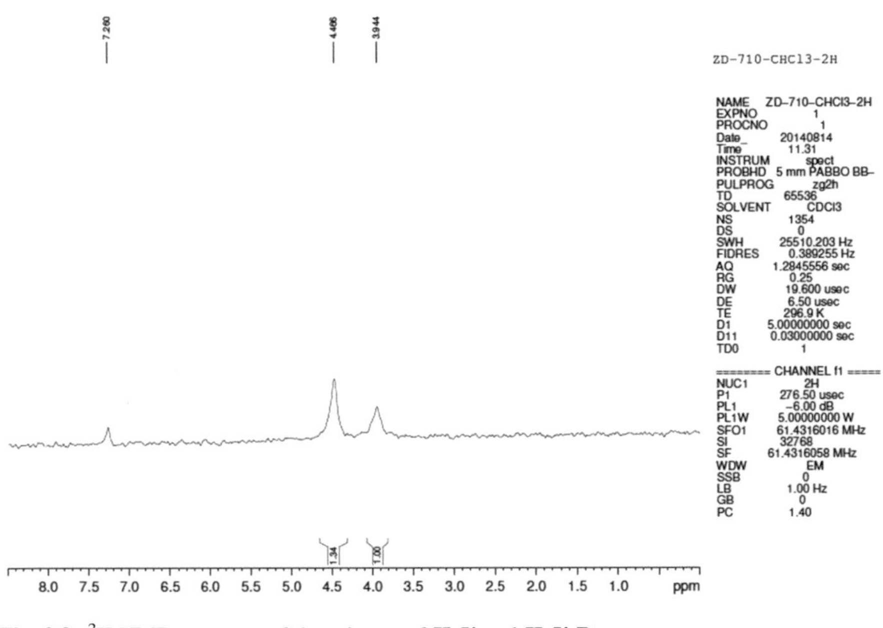

Fig. 2.9 ^{2}H NMR spectrum of the mixture of **II-5j** and **II-5j-D$_2$**

Starting from dimeric species [1-I-2-Li-o-$C_2B_{10}H_{12}$]$_2$ (**II-INT1**), three reaction pathways (**Path A**, **Path B**, and **Path C**) leading to the formation of α-carboranylation product were located. **Path A** involves one carborane molecule-assisted dissociation of LiI via transition state **II-TS2S** (21.6 kcal/mol) to give a singlet carboryne **II-CBRS** (11.6 kcal/mol). The complexation of carboryne **II-CBRS** with NMe_3 generates a zwitterionic intermediate **II-INT3** (4.1 kcal/mol). From

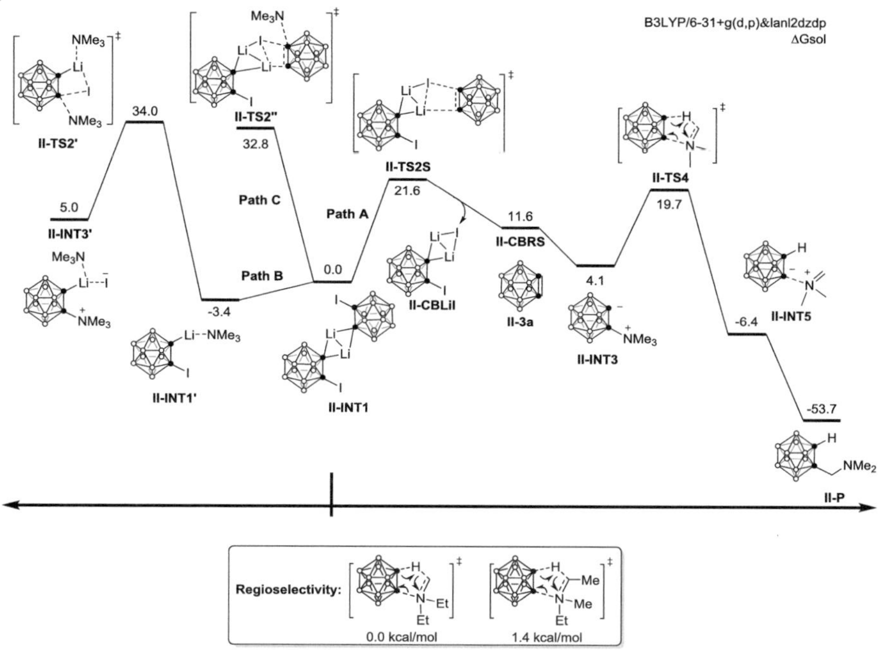

Fig. 2.10 Energy profile calculated for the reaction pathway

II-INT3, a proton transfer from the α-position of the amine to cage carbon produces another zwitterionic intermediate **II-INT5** possessing a carborane anion and an iminium ion. The nucleophilic addition of carborane anion to iminium ion leads to the formation of the α-carboranylation product **II-P**. The transition state related to the nucleophilic addition has not been located, whereas we assume that this step is very facile due to the charge separation of carborane anion and iminium ion.

On the other hand, **Path B** involving dissociation of the dimeric species [1-I-2-Li-o-$C_2B_{10}H_{12}$]$_2$ and nucleophilic attack of NMe_3 to the C–I bond was calculated. The results show that **Path B** is less favorable than **Path A** due to the higher barrier of the nucleophilic substitution (**II-TS2′**: 34.0 kcal/mol vs. **II-TS2S**: 21.6 kcal/mol). Similarly, the calculated results show that **Path C** involving direct nucleophilic attack of NMe_3 to the C–I bond of **II-INT1** is also less favorable than **Path A** due to the high energy of the nucleophilic substitution in transition state **II-TS2″** (**II-TS2″**: 32.8 kcal/mol vs. **II-TS2S**: 21.6 kcal/mol).

For the most favorable reaction pathway **Path A**, the calculated rate-determining step is the formation of carboryne **II-CBRS** via **II-TS2S**, other than the proton shift process via **II-TS4**, which is consistent with the experimental observations. In line with the formation of a singlet carboryne **II-CBRS** (11.6 kcal/mol) in **II-Path A**, the formation of a triplet carboryne **II-CBRT** was also calculated and the calculated results reveal that the formation of triplet carboryne **II-CBRT** is disfavored, due to the high energy of **II-CBRT** (**II-CBRT**: 31.6 kcal/mol vs. **II-TS2S**: 21.6 kcal/mol).

This result is also consistent with the experimental observation that radical process is not involved in this reaction.

Experimentally, *N,N*-dimethylbutylamine **II-4g** or *N*-methyldibutylamine **II-4h** underwent *α*-carboranylation with >14:1 or 1.8:1 selectivity for methyl C–H bond cleavage over secondary C–H bond, reflecting an inherent >4.7:1 or ~2.4:1 selectivity after considering the number of 1 and 2° C–H bonds presented in **II-4g** or **II-4h**. Consistently, the calculated results show that the H shift from methyl group is more favorable than that from ethyl group by 1.4 kcal/mol (Fig. 2.10). We assume that the preference for primary C–H bond over secondary C–H bond is caused by the developing negative charge at *α* carbon, and primary carbon anion is more stable than the secondary carbon anion.

On the basis of both experimental and theoretical results, a plausible mechanism is proposed in Scheme 2.5. Elimination of LiI from **II-2a** gives the reactive intermediate *o*-carboryne **II-3a**. Nucleophilic attack of tertiary amine **II-4** generates the zwitterionic intermediate **II-A** (path a). Alternatively, a S_N2-type nucleophilic substitution at the cage C–I bond and elimination of LiI can also lead to the formation of **II-A** (path b). **II-A** undergoes an intramolecular hydrogen transfer to afford an iminium ion salt **II-B**. Intramolecular addition yields the final product **II-5**. However, isolation of the proposed *N*-carboranyl ammonium salt intermediate **II-A** failed. The computational results suggest that Path A is more favorable than Path B. The calculations also reveal that the rate-determining step for Path A is the formation of *o*-carboryne, and the hydrogen shift is a fast process. The nature of the hydrogen transfer is likely to be a proton transfer, as indicated by DFT calculations. In this regard, α-carboranylation at benzylic position is dominant over that of aliphatic position (Table 2.2, entries 9–12) since benzylic proton is much acidic than aliphatic one.

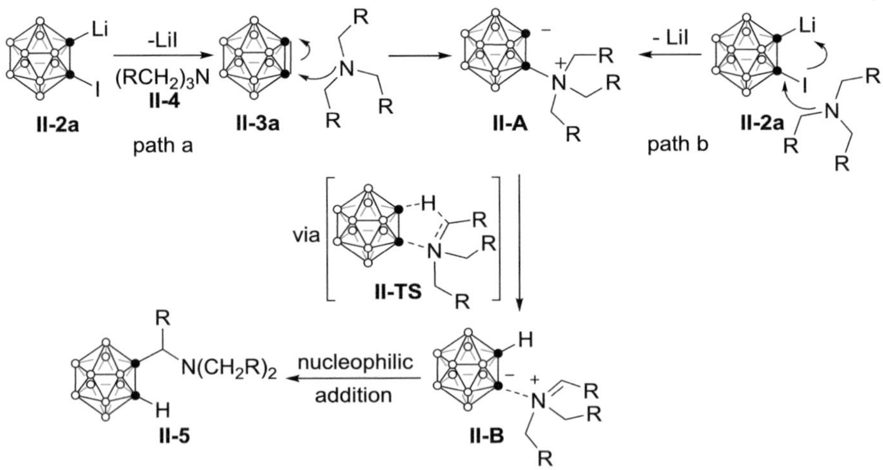

Scheme 2.5 Proposed pathways for α-carboranylation of amines

2.5 Summary

We have described the regioselective insertion of *o*-carborynes into *α*-C–H bonds of tertiary amines. In this process, the more electron-poor C–H bonds are favored. The bonding form of *o*-carboryne intermediate is involved in this reaction, as suggested by mechanistic studies as well as DFT calculations. This work represents another unique reactivity pattern of *o*-carboryne, which also serves as an efficient approach to 1-aminoalkyl carboranes that might find applications in catalysis and medicine.

References

1. Corey EJ, Cheng XM (1989) The logic of chemical synthesis. John Wiely & Sons, New York
2. (a) Lawrence SA (2004) Amines: synthesis, properties and applications. Cambridge University Press, New York (b) Nugent TC (2010) Chiral amine synthesis: methods, development and applications. Wiley-VCH Verlag, Weinheim
3. (a) Beak P, Basu A, Gallagher DJ et al (1996) Regioselective, diastereoselective, and enantioselective lithiation-substitution sequences: reaction pathways and synthetic applications. Acc Chem Res 29(11):552–560 (b) Hoppe D, Hense T (1997) Enantioselective synthesis with lithium/(-)-sparteine carbanion pairs. Angew Chem Int Ed 36(21):2283–2316
4. (a) Campos KR (2007) Direct sp(3) C–H bond activation adjacent to nitrogen in heterocycles. Chem Soc Rev 36(7):1069–1084 (b) Mitchell EA, Peschiulli A, Lefevre N et al (2012) Direct alpha-functionalization of saturated cyclic amines. Chem-Eur J 18(33):10092–10142
5. (a) Crabtree RH (2005) The organometallic chemistry of transition metals, 4th edn. Wiley Interscience, New York (b) McQuillin FJ, Parker, DG, Stephenson GR (1991) Transition metal organometallics for organic synthesis. Cambridge University Press, Cambridge (c) Tsuji J (2002) Transition metal reagents and catalysts: innovations in organic synthesis. Wiley, Chichester (d) Li CJ (2009) Cross-dehydrogenative coupling (cdc): exploring C–C bond formations beyond functional group transformations. Acc Chem Res 42(2):335–344 (e) Murahashi SI, Zhang D (2008) Ruthenium catalyzed biomimetic oxidation in organic synthesis inspired by cytochrome p-450. Chem Soc Rev 37(8):1490–1501
6. (a) Yeung CS, Dong VM (2011) Catalytic dehydrogenative cross-coupling: forming carbon-carbon bonds by oxidizing two carbon-hydrogen bonds. Chem Rev 111(3):1215–1292 (b) Allen JM, Lambert TH (2011) Tropylium ion mediated alpha-cyanation of amines. J Am Chem Soc 133(5):1260–1262 (c) Schnurch M, Dastbaravardeh N, Ghobrial M et al (2011) Functionalization of saturated and unsaturated heterocycles via transition metal catalyzed c-h activation reactions. Curr Org Chem 15(15):2694–2730 (d) Li CJ, Li ZP (2006) Green chemistry: the development of cross-dehydrogenative coupling (cdc) for chemical synthesis. Pure Appl Chem 78(5):935–945 (e) Murahashi SI (1995) Synthetic aspects of metal-catalyzed oxidations of amines and related reactions. Angew Chem-Int Ed Engl 34 (22):2443–2465 (f) Murahashi SI, Naota T, Kuwabara T et al (1990) Ruthenium-catalyzed oxidation of amides and lactams with peroxides. J Am Chem Soc 112(21):7820–7822
7. (a) Hao E, Vicente MGH (2005) Expeditious synthesis of porphyrin-cobaltacarborane conjugates. Chem Comm (10):1306–1308 (b) Soloway AH, Tjarks W, Barnum BA et al (1998) The chemistry of neutron capture therapy. Chem Rev 98(4):1515–1562 (c) Hawthorne MF, Maderna A (1999) Applications of radiolabeled boron clusters to the diagnosis and treatment of cancer. Chem Rev 99(12):3421–3434 (d) Valliant JF, Guenther KJ,

King AS et al (2002) The medicinal chemistry of carboranes. Coord Chem Rev 232(1–2): 173–230

8. (a) Hosmane NS, Maguire JA (2003) Recent advances in the chemistry of metallacarboranes. Eur J Inorg Chem (22):3989–3999 (b) Xie Z (2006) Group 4 metallocenes incorporating constrained-geometry carboranyl ligands. Coordin Chem Rev 250(1–2):259–272 (c) Hosmane NS, Zhu Y, Maguire JA et al (2009) Nano and dendritic structured carboranes and metallacarboranes: From materials to cancer therapy. J Organomet Chem 694(11):1690–1697 (d) Shen H, Chan HS, Xie Z (2006) Guanylation of amines catalyzed by a half-sandwich titanacarborane amide complex. Organometallics 25(23):5515–5517 (e) Shen H, Xie ZW (2008) Titanacarborane amide catalyzed transamination of guanidines. Organometallics 27 (12):2685–2687
9. (a) Barth RF (2003) A critical assessment of boron neutron capture therapy: An overview. J Neuro-Oncol 62(1):1–5 (b) Hawthorne MF, Lee MW (2003) A critical assessment of boron target compounds for boron neutron capture therapy. J Neuro-Oncol 62(1):33–45
10. (a) Lee YJ, Lee JD, Jeong HJ et al (2005) Sterically protected titanium (aminoethyl) dicarbollides: synthesis of novel constrained-geometry complexes showing an unusual cage b, n-cyclization. Organometallics 24(12):3008–3019 (b) Shen H, Xie Z (2009) Titanacarborane mediated c-n bond forming/breaking reactions. J Organomet Chem 694(11):1652–1657 (c) Lee JD, Lee YJ, Son KC et al (2007) New types of constrained geometry group 4 metal complexes derived from the aminomethyldicarbollyl ligand system: synthesis and structural characterization of mono-dicarbollylamino and bis-dicarbollylamino group 4 metal complexes. Organometallics 26(14):3374–3384 (d) Shen H, Chan HS, Xie Z (2007) Synthesis, structure, and reactivity of [σ:η1:η5-$(OCH_2)(Me_2NCH_2)$ $C_2B_9H_9$] Ti (NR_2)(R = Me, Et). Organometallics 26(10): 2694–2704
11. (a) Terrasson V, Planas JG, Prim D et al (2009) General access to aminobenzyl-o-carboranes as a new class of carborane derivatives: entry to enantiopure carborane-amine combinations. Chem-Eur J 15(44):12030–12042 (b) Li YQ, Carroll PJ, Sneddon LG (2008) Ionic-liquid-promoted decaborane dehydrogenative alkyne-insertion reactions: a new route to o-carboranes. Inorg Chem 47(20):9193–9202 (c) Maderna A, Herzog A, Knobler CB et al (2001) The syntheses of amphiphilic camouflaged carboranes as modules for supramolecular construction. J Am Chem Soc 123(42):10423–10424 (d) Wu Y, Carroll PJ, Kang SO et al (1997) Synthesis, characterization, and reactivity of isocyanato dicarbaboranes obtained from o-carborane. Inorg Chem 36(21):4753–4761 (e) Olid D, Nunez R, Vinas C et al (2013) Methods to produce B-C, B-P, B-N and B-S bonds in boron clusters. Chem Soc Rev 42(8): 3318–3336
12. (a) Qiu Z, Wang SR, Xie Z (2010) Nickel-catalyzed regioselective [2 + 2 + 2] cycloaddition of carboryne with alkynes. Angew Chem Int Edit 49(27):4649–4652 (b) Qiu Z, Xie Z (2014) Generation and reactivity of o-carborynes. Dalton Trans 43(13):4925–4934
13. Wang SR, Qiu Z, Xie Z (2011) Regioselective insertion of carborynes into ethereal C-H bond: facile synthesis of alpha-carboranylated ethers. J Am Chem Soc 133(15):5760–5763
14. (a) Cottrell TL (1958) The strengths of chemical bonds, 2nd edn. Butterworth, London (b) Darwent BdeB. National standard reference data series. National Bureau of Standards, Washington (c) Blanksby SJ, Ellison GB (2003) Bond dissociation energies of organic molecules. Acc Chem Res 36(4):255–263
15. (a) Wittig G, Behnisch W (1958) Dehydrobenzol und n-methyl-pyrrol. Chem Ber 91(11): 2358–2365 (b) Wolthuis E, Cady W, Roon R et al (1966) Preparation and properties of 1,4-dihydronaphthalen-1,4-imines. J Org Chem 31(6):2009–2011 (c) Lepley AR, Becker RH, Giumanin AG (1971) Benzyne addition to n,n-dimethylbenzylamine. J Org Chem 36(9): 1222–1227 (d) Sato Y, Ban Y, Aoyama T et al (1976) Reaction of N,N-dimethyl-2-triorganosilylethylamines with benzyne. J Org Chem 41(11):1962–1965 (e) Sato Y, Toyooka T, Aoyama T et al (1976) Reaction of N-(triorganosilylmethyl)dialkylamines with benzyne. J Org Chem 41(22):3559–3564 (f) Cant AA, Bertrand GHV, Henderson JL et al (2009) The benzyne aza-claisen reaction. Angew Chem Int Ed 48(28):5199–5202

(g) Bhojgude SS, Kaicharla T, Biju AT (2013) Employing arynes in transition-metal-free monoarylation of aromatic tertiary amines. Org Lett 15(21):5452–5455

16. Xu X, Li X, Ma L et al (2008) An unexpected diethyl azodicarboxylate-promoted dehydrogenation of tertiaryamine and tandem reaction with sulfonyl azide. J Am Chem Soc 130(43):14048–14049
17. Frisch MJ, Trucks GW, Schlegel HB et al (2009) Gaussian 09, revision D. 01. Gaussian, Inc., Pittsburgh
18. (a) Becke AD (1993) Density-functional thermochemistry. 3. The role of exact exchange. J Chem Phys 98(7):5648–5652 (b) Lee CT, Yang WT, Parr RG (1988) Development of the colle-salvetti correlation-energy formula into a functional of the electron-density. Phys Rev B 37(2):785–789
19. Check CE, Faust TO, Bailey JM et al (2001) Addition of polarization and diffuse functions to the LANL2DZ basis set for p-block elements. J Phys Chem A 105(34):8111–8116
20. (a) Schlitter J, Klahn M (2003) The free energy of a reaction coordinate at multiple constraints: A concise formulation. Mol Phys 101(23–24):3439–3443 (b) Fukui K (1981) The path of chemical-reactions—the irc approach. Acc Chem Res 14(12):363–368
21. (a) Cossi M, Rega N, Scalmani G et al (2003) Energies, structures, and electronic properties of molecules in solution with the c-pcm solvation model. J Comput Chem 24(6):669–681 (b) Cossi M, Barone V, Cammi R et al (1996) Ab initio study of solvated molecules: a new implementation of the polarizable continuum model. Chem Phys Lett 255(4–6):327–335 (c) Cossi M, Barone V, Mennucci B et al (1998) Ab initio study of ionic solutions by a polarizable continuum dielectric model. Chem Phys Lett 286(3–4):253–260 (d) Cossi M, Barone V, Robb MA (1999) A direct procedure for the evaluation of solvent effects in mc-scf calculations. J Chem Phys 111(12):5295–5302

Chapter 3
Synthesis of Carborane-Functionalized Heterocycles: Dearomative [2 + 2] Cycloaddition and sp^2 C–H Insertion Reaction

Abstract This chapter focuses on reaction of *o*-carboryne with *N*-protected indoles. An unexpected dearomative [2 + 2] cycloaddition is observed in the case of *N*-TMS indoles producing *o*-carborane-fused indolines in excellent chemoselectivity and isolated yields. In addition to the [2 + 2] cycloadducts, carboranylindoles derived from formal insertion reaction are also obtained for *N*-aryl indoles in which the product ratio is dependent upon the nature of substituents on aryl rings. The formal insertion reaction is dominant in the reaction of *N*-alkyl indoles. These protocols offer general and efficient methods for the preparation of *o*-carborane-functionalized indoles and indolines as well as other heterocycles. The observed dearomative [2 + 2] cycloaddition also represents the first example for indoles to undergo such reaction in the absence of transition metals or without UV irradiation.

3.1 Introduction

Dearomatization reactions are one of the important transformations of aromatic molecules as they lead directly to various cyclic or heterocyclic skeletons from structurally simple substrates [1]. In particular, due to the potent biological activities and pharmaceutical applications of indoline alkaloids, dearomatization of indoles has long been subject of interest in synthetic chemistry [1, 2]. Among those elegant dearomative strategies, dearomative cycloaddition of the indole derivatives represents an attractive, straightforward as well as atom-economic approach to indoline compounds [3]. It has been shown that indole can undergo [3 + 2], [3b–e, 4] [4 + 2] [3f–I, 5], and [5 + 2] [6] cycloaddition reactions to generate cyclopenta[b]indoles, hydrocarbazoles, and cyclohepta[b]indoles, respectively (Fig. 3.1).

Despite great efforts in the dearomatization of indoles, only three reports are available on non-photo-induced dearomative [2 + 2] cycloaddition of indoles (Scheme 3.1) [7, 8]. Zhang first developed an intramolecular gold-catalyzed [2 + 2] cycloaddition of indoles [8a]. Mascareñas and López reported an intermolecular [2 + 2] cycloaddition of indole but it worked with only one indole substrate and the

D. Zhao, *Functionalization of Carborane via Carboryne Intermediates*,
Springer Theses, DOI 10.1007/978-981-10-1569-4_3

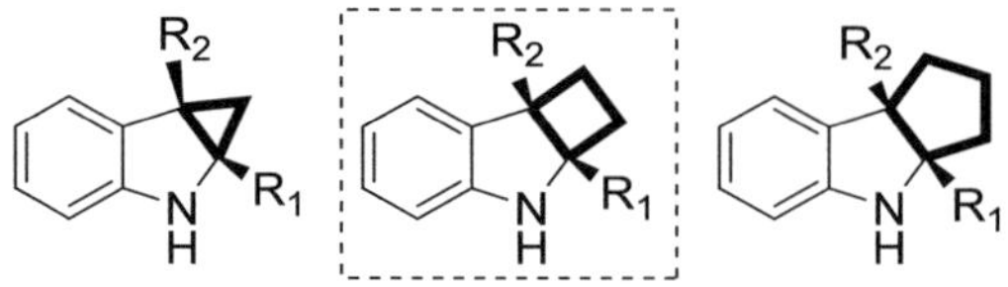

Fig. 3.1 Examples of 2,3-fused indolenyl scaffolds

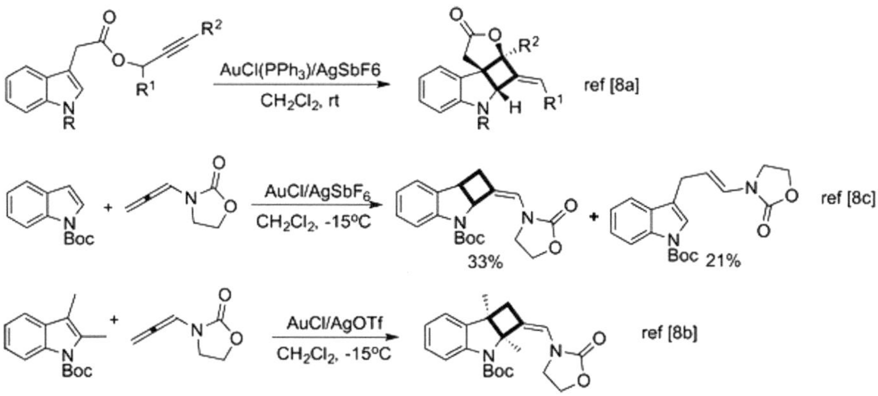

Scheme 3.1 Transition-metal-catalyzed [2 + 2] dearomative cycloaddition of indoles

chemoselectivity is poor [8c]. Very recently, Bandini and coworkers expanded this reaction, which was first reported by Mascareñas and López, to an enantioselective fashion [8b]. However, it only worked with the 2,3-disubstituted indole. To the best of our knowledge, transition-metal-free protocols to address cyclobuta-fused indoline frameworks are still absent in the literature.

On the other hand, direct C–H functionalization of heterocycles has received great attention due to the structural prevalence of substituted heterocycles in drugs, natural products, and other biologically active molecules [9]. In this regard, indole derivatives represent a system of particular interest and importance [10]. As a result, numerous efforts have been directed toward the development of mild, selective, and efficient methods for the functionalization of indole molecules, especially with regard to transition-metal-catalyzed [11] C–H bond activation and direct Friedel–Crafts alkylations as well as allylic alkylations [11a, d, 12].

In view of a growing interest in the applications of carborane-containing molecules in medicine [13], materials science [14], and organometallic/coordination chemistry [15], the development of new methods for the synthesis of such molecules has received considerable attention. For instance, carborane derivatives bearing π–substituents (aryl and heteroaryl) have shown promising photophysical properties as light-emitting materials [16]. Additionally, carboranes that featuring a

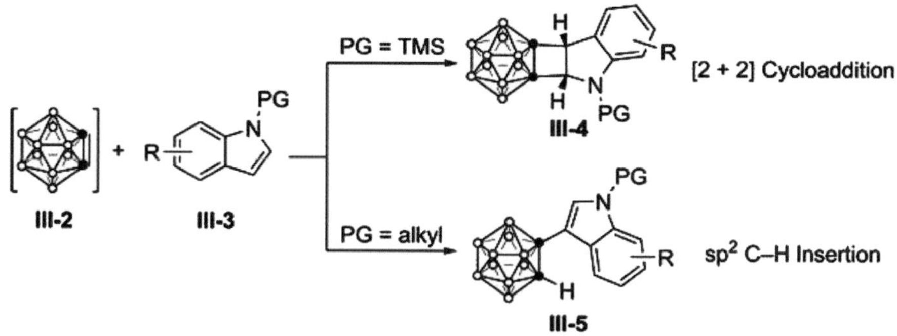

Scheme 3.2 Reaction of *o*-carboryne with indoles

5,6,7-trimethoxyindole unit are attractive DNA-binding boron sources for boron neutron capture therapy (BNCT) [17].

Considering the unique properties of indoles and carboranes, carborane-functionalized indole or indolines may find applications in medicine and materials. Surprisingly, despite the remarkable progress in carborane chemistry, straightforward and general synthesis of carborane-functionalized heterocycles such as indolinyl- or indolyl-*o*-carboranes still represents a very challenging task [18, 19]. During our ongoing research on the functionalization of carborane via *o*-carboryne intermediate [20], we found that *o*-carboryne (**III-2**) reacted readily with *N*-protected indoles in two reaction pathways, in which the chemoselectivity of the reaction depended on the protecting group of the indole substrates (Scheme 3.2). In this section, we report this new methodology, which provides an efficient, chemoselective, and highly controllable synthesis of carboranyl-substituted indoles and indolines.

3.2 Reaction of *o*-Carboryne with *N*-TMS Indoles

Our investigations began with the reaction of *o*-carboryne with *N*-TMS indole, and the results are compiled in Table 3.1. Treatment of 1-Br-2-Li-*o*-$C_2B_{10}H_{10}$ (**III-1**) (prepared in situ by mixing a 1:1 molar ratio of 1,2-Li_2-*o*-$C_2B_{10}H_{10}$ with 1,2-Br_2-*o*-$C_2B_{10}H_{10}$ in *n*-hexane) with 5 equiv of 1-TMS-1*H*-indole (**III-3aa**) in *n*-hexane at 60 °C for 6 h gave [2 + 2] cycloadduct **III-4aa** in 81 % isolated yield (Table 3.1, entry 1). Lowering the amount of **III-3aa** to 2.0 and 1.1 equiv showed negligible influence in the isolated yield of **III-4aa** (Table 3.1, entries 1–3). Increasing the reaction temperature to 80 °C did not improve the yield (Table 3.1, entry 4). However, heating was necessary for this reaction as only a trace amount of **III-4aa** was formed after 24 h at room temperature (Table 3.1, entry 6).

Table 3.1 Optimization of reaction conditions[a]

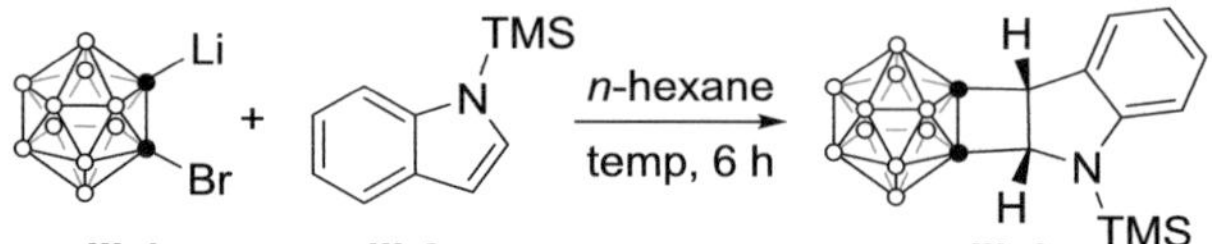

Entry	**III-3aa** (equiv)	Temp (°C)	**III-4aa** (%)[b]
1	5.0	60	81
2	2.0	60	87
3	1.1	60	86
4	1.1	80	78
5	1.1	40	38
6[c]	1.1	25	trace

[a]Reaction conditions: **III-1** (1.0 mmol), indole **III-3a** (1.1, 2.0 or 5.0 mmol) in *n*-hexane (10 mL), 60 °C, 6 h, *TMS* Trimethylsilyl
[b]Isolated products are given
[c]Stirred for 24 h

Under the optimal reaction conditions shown in Table 3.1, entry 3, a broad array of *N*-TMS-substituted indoles underwent [2 + 2] cycloaddition with *o*-carboryne, producing carboranyl-fused indolines in 61–92 % yields (Table 3.2). For instance, the C2,C3-carboranyl-fused indolines **III-4ab** and **III-4ac** were obtained in good yields for 2-phenyl and 2-Me-substituted *N*-TMS-1*H*-indole **III-3ab** and **III-3ac** (Table 3.2, entries 2–3). For indole **III-3ad** featuring a methyl group at C3 position, in addition to the [2 + 2] cycloadduct **III-4ad**, an ene product [21] was isolated in 24 % isolated yield (Table 3.2, entry 4). In contrast, only the [2 + 2] cycloadduct was obtained in 78 % isolated yield and the formation of the undesired ene product was suppressed for C2-,C3-disubstituted indole **III-3ae** (Table 3.2, entry 5). Electron-withdrawing (F, Cl, Br, Ph) as well as electron-donating (Me, OMe, *i*Pr) benzenoid substituents were well tolerated in this reaction (Table 3.2, entries 6–14). The molecular structures of **III-4aa**, **III-4ab**, **III-4ad**, **III-4ag**, and **III-4am** were further confirmed by single-crystal X-ray analyses (Figs. 3.2, 3.3, 3.4, 3.5 and 3.6).

3.3 Effects of *N*-Substituents

Interestingly, as shown in Table 3.3, the electronic and steric properties of the *N*-substituents have a dramatic influence in controlling the chemoselectivity of the reactions. For instance, a significant steric effect was observed for *N*-silyl indoles. *N*-(*t*-butyldimethylsilyl) indole **III-3b** gave both [2 + 2] cycloadduct **III-4b** and formal C–H insertion product [22, 23] **III-5b** in 33 and 42 % yields, respectively.

Table 3.2 Substrate scope for [2 + 2] cycloaddition of *N*-TMS indoles[a]

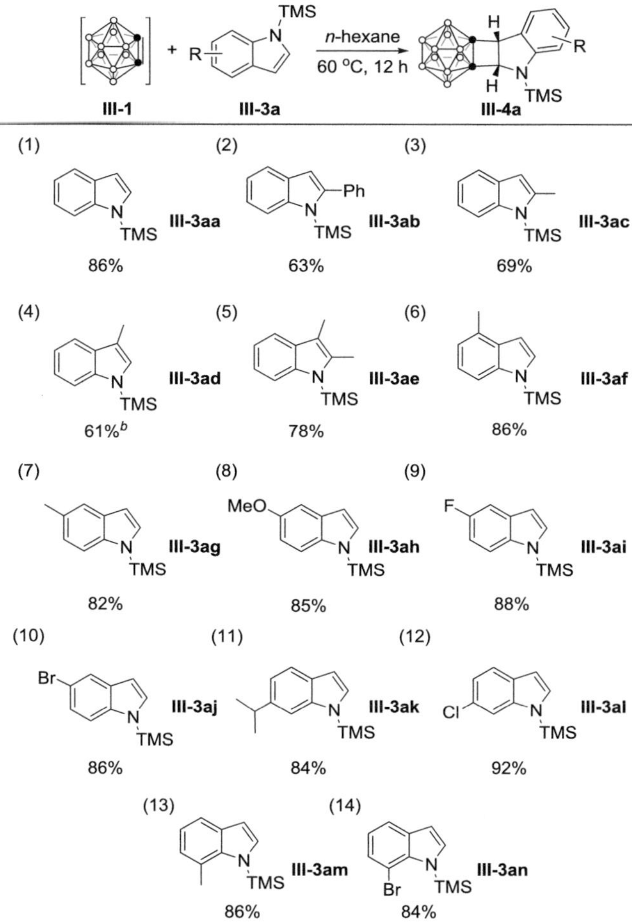

[a]Reaction conditions: **III-1** (1.0 mmol), indole **III-3a** (1.1 mmol) in *n*-hexane (10 mL), 60 °C, 6 h. Yields of isolated products were given

[b]Ene reaction product was also isolated in 24 % yield (see Scheme 3.5, Eq. 3.5)

The [2 + 2] cycloaddition was completely inhibited for a more bulky triisopropylsilyl (TIPS) group (Table 3.3, entries 2–3). For unprotected 1*H*-indole **III-3d**, an equimolar reaction gave only acid–base reaction product, 1-Br-*o*-$C_2B_{10}H_{11}$ (p*K*a values: 21.0 for **III-3d** vs. ~23 for *o*-carborane) [10c, 18]. The desired 3-carboranyl-1*H*-indole **III-5d** was isolated in 30 % yield (60 % based on recovered starting material) when 0.5 equiv of **III-3d** was used (Table 3.3, entry 4). The corresponding formal insertion products were obtained in very good yields with

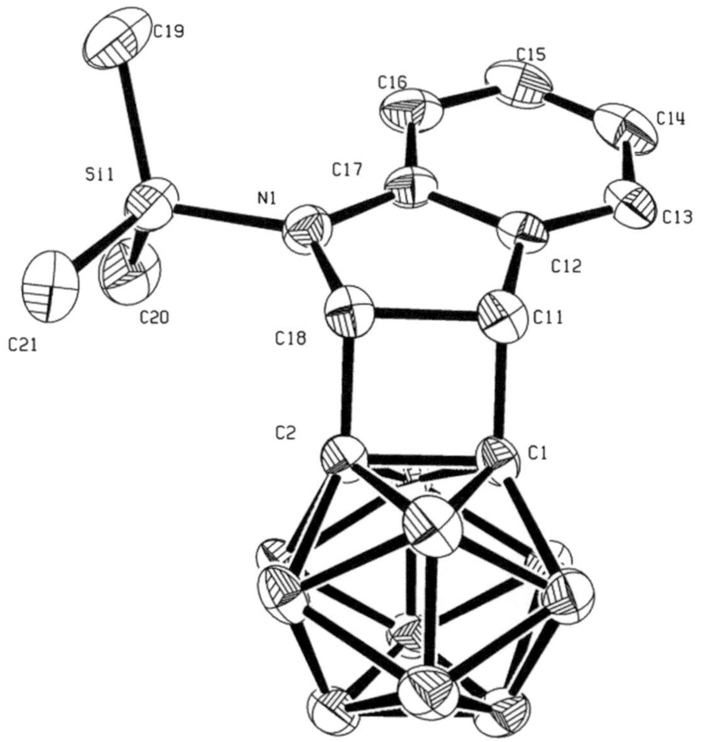

Fig. 3.2 Molecular structure of **III-4aa**

excellent regioselectivity for *N*-alkyl indoles (alkyl = methyl, allyl, benzyl) (Table 3.3, entries 5–7). For *N*-aryl indoles, both the formal insertion product and the [2 + 2] cycloadduct were formed (Table 3.3, entries 8–10). A fluoro-substituent at the para position of the phenyl ring increased the molar ratio of [2 + 2] cycloadduct over the C–H insertion product to 1.4, while a 4-methoxy-substituent decreased such ratio to 0.7. The above results indicate that the [2 + 2] cycloaddition reaction is more favored if the protecting group on the nitrogen is electron-deficient, which may shed some light on the reaction mechanism (vide infra). Many attempts to utilize other *N*-protected (e.g., protecting group = $SO_2C_6H_4CH_3$, $CO_2(CH_3)_3$, $COCH_3$) indoles failed due probably to the side reactions of the protecting group with precursor **III-1**. These results indicate the synthetic potential of these two reactions as carboranyl-substituted indoles and indolines that are unable to be constructed before can now be synthesized in an efficient and highly controlled manner by utilizing different *N*-protecting groups (alkyl or TMS). The molecular structures of **III-5ea**, **III-4h**, and **III-4j** were further confirmed by single-crystal X-ray analyses (Figs. 3.7, 3.8, and 3.9).

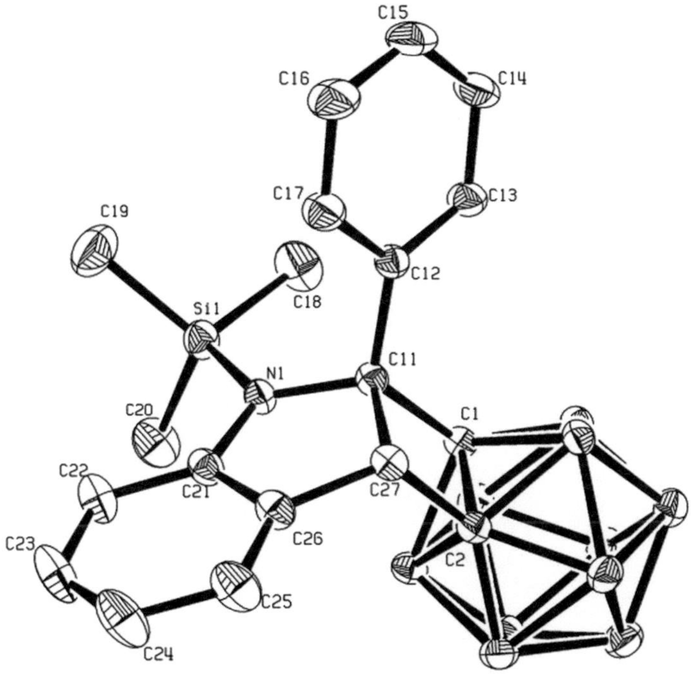

Fig. 3.3 Molecular structure of **III-4ab**

3.4 Reaction of *o*-Carborynes with *N*-Methyl Indoles

We then investigated the substrate scope with a variety of *N*-methyl-protected indole derivatives for the synthesis of carborane-substituted indoles and the results are compiled in Table 3.4. Both electron-withdrawing (F, Cl, Br) and electron-donating (Me, iPr, OMe) groups on the indole ring were well tolerated, producing the corresponding C3-carboranylindoles **III-4ea–4ek** in 60–91 % yields. Notably, the halogen substituent can be used for further synthetic elaboration.

3.5 Reaction of *o*-Carboryne with Other *N*-Heterocycles

This reaction is also compatible with other *N*-heterocycles, providing a general and straightforward approach for carboranyl heterocycles (Table 3.5). It is noted that the construction of such molecules often involves multi-step synthesis [24]. Reactions of 1-methyl-1*H*-benzimidazole **III-6a**, benzoxazole **III-6b** and benzothiazole **III-6c** afforded the corresponding C2-carboranylated heterocycles in moderately isolated yields (Table 3.5, entries 1–3). Both C2 and C3 carboranylated products with a molar ratio of 1:1.5 were isolated in 80 % yield for *N*-methyl-1*H*-

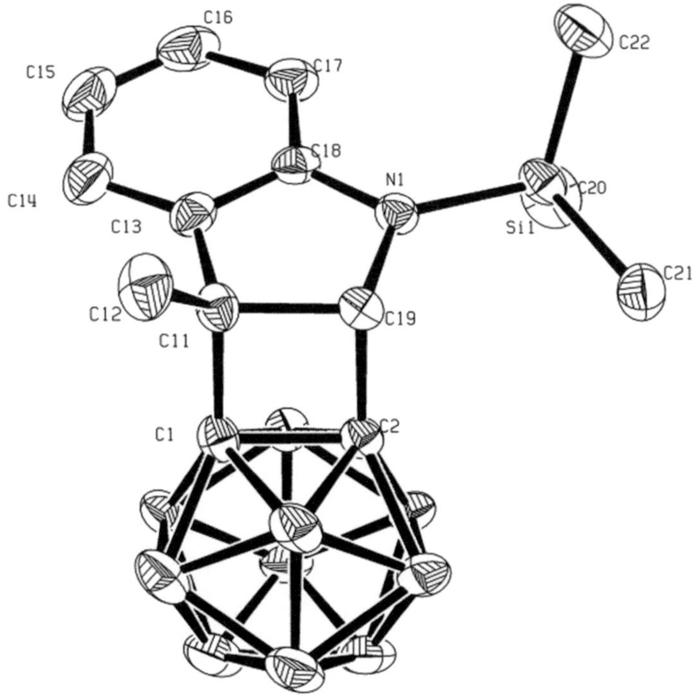

Fig. 3.4 Molecular structure of **III-4ad**

pyrrole **III-6d** (Table 3.5, entry 4). The reaction of 4,5-dimethylthiazole **III-6e** proceeded smoothly to give C2-carboranylated product in 87 % yield, whereas quinoline **III-6f** was a poor substrate as only 40 % of the desired product was obtained (Table 3.5, entries 5–6). The molecular structures of **III-7c**, **III-7da,** and **III-7f** were further confirmed by single-crystal X-ray analyses (Figs. 3.10, 3.11 and 3.12). The molecular structures of **III-7c**, **III-7da** and **III-7f** were further confirmed by single-crystal X-ray analyses (Figs. 3.10, 3.11 and 3.12).

3.6 Transformation of [2 + 2] Cycloadducts

We also examined the transformations of the [2 + 2] cycloadducts. As shown in Scheme 3.3, the [2 + 2] cycloadduct **III-4j** rearomatizes to **III-5j** under thermal or acidic conditions. For methoxyl-substituted cycloadduct **III-4ah**, this ring-opening process occurred more easily to give the rearomatized product (ring-opening product) upon treatment with silica gel. On the other hand, treatment of **III-4ah** or

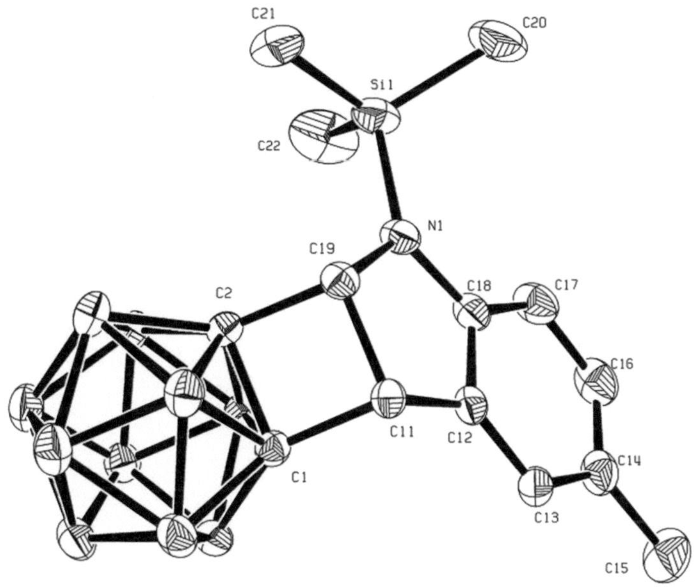

Fig. 3.5 Molecular structure of **III-4ag**

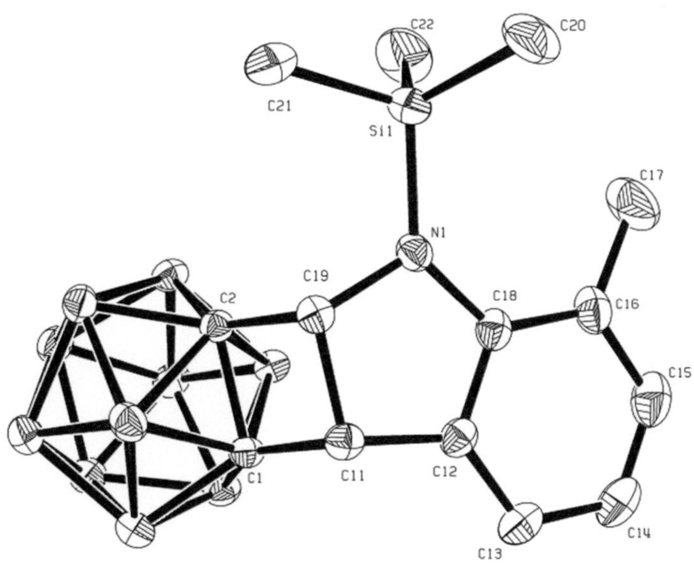

Fig. 3.6 Molecular structure of **III-4am**

Table 3.3 Effects of *N*-substituents[a]

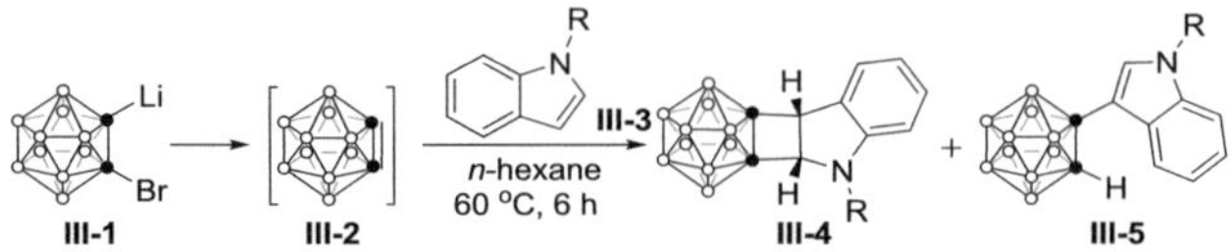

Entry	R	III-4 (%)[b]	III-5 (%)[b]
1	Me_3Si (**III-3aa**)	86	–
2	$tBuMe_2Si$ (**III-3b**)	33	42
3	iPr_3Si (**III-3c**)	–	83
4[c]	H (**III-3d**)	–	30
5	Me (**III-3ea**)	–	85
6	allyl (**III-3f**)	–	85
7	$C_6H_5CH_2$ (**III-3 g**)	–	78
8	Ph (**III-3 h**)	39	38
9	$4\text{-}MeOC_6H_4$ (**III-3i**)	30	45
10	$4\text{-}FC_6H_4$ (III-3j)	43	30

[a]Reaction conditions: **III-1** (1.0 mmol), indole **III-3** (1.1 mmol) in *n*-hexane (10 mL), 60 °C, 6 h
[b]Isolated yields
[c]0.5 equiv of 3d was used

III-4ac with KF afforded the rearomatized product **III-5dh** or **III-5dc** in quantitative yield. For 3-methyl-substituted indoline **III-4ae**, no rearomatization product was observed; instead, a carborane-substituted indolenine **III-8** was generated in quantitative yield upon desilylation. The molecular structures of **III-5ah** and **III-8** were further confirmed by single-crystal X-ray analyses (Figs. 3.13 and 3.14).

The aforementioned results show that the formation of 1*H*-indoles or indolenine is dependent on the C3-substituent of the indoline molecule (Scheme 3.4). When the substituent is hydrogen, carboranyl-1*H*-indole is produced through a ring-opening reaction followed by intramolecular hydrogen transfer sequence. For C3-substituted indoline, desilylation affords the corresponding carboranyl-substituted indolenine as it cannot undergo rearomatization process. These results also suggest that the ring-opening or rearomatized product is favored when the electron density of the indole nitrogen is increased.

3.7 Mechanistic Study

To gain some insights into the reaction mechanism, several control experiments were conducted (Scheme 3.5). Neither the [2 + 2] cycloadduct **III-4aa** nor the insertion product **III-5ea** was detected from the reaction of 1,2-dilithio-*o*-carborane (**III-9**) with **III-3aa** or **III-3ea** (Eq. 3.1). Furthermore, no reaction occurred when **III-3ea** was added to a 1:1 molar ratio of 1-bromo-2-methyl-*o*-carborane (**III-10**) and 1-lithio-2-methyl-*o*-carborane (**III-11**) (Eq. 3.2). The above experiments may

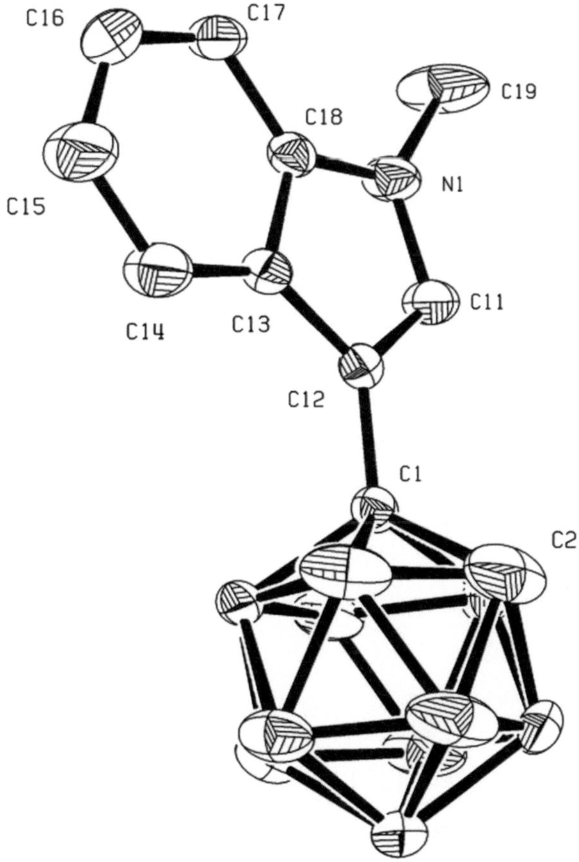

Fig. 3.7 Molecular structure of **III-5ea**

indicate the intermediacy of *o*-carboryne in these reactions. The desired product **III-4aa** or **III-5ea** was still formed in 80 or 75 % isolated yield in the presence of 1.1 equiv of a radical scavenger, 2,2,6,6-tetramethylpiperidine-1-oxyl (TEMPO), which rules out the involvement of any radical species (Eq. 3.3). When the C3 position of indole derivative was blocked, for example, **III-3el**, the expected C2-carboranyl-substituted product **III-4el** was not observed in the crude reaction mixture. Instead, the sp^3 C–H insertion product **III-13** and ene reaction product **III-14** were produced in 50 and 25 % isolated yields. The latter (**14**) was then converted quantitatively to **III-5el** in $CDCl_3$ at room temperature after 6 h (Eq. 3.4). These results not only indicate the involvement of *o*-carboryne intermediate in this reaction, but also clarify that the C3 position of the indole ring is the initial reaction site for the formation of the sp^2 C–H insertion product. On the other hand, the [2 + 2] cycloaddition was still the dominant pathway in the reaction of 1-TMS-3-methyl-1*H*-indole **III-3ad**, although minor ene adduct **III-20** was also observed (Eq. 3.5).

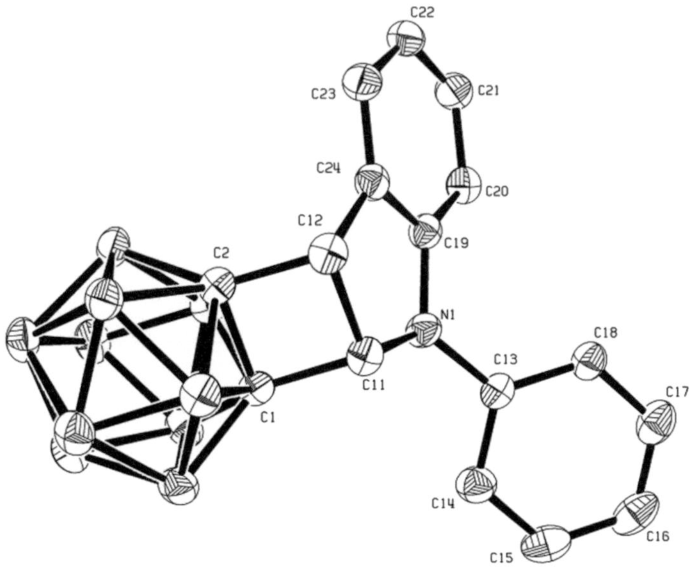

Fig. 3.8 Molecular structure of **III-4h**

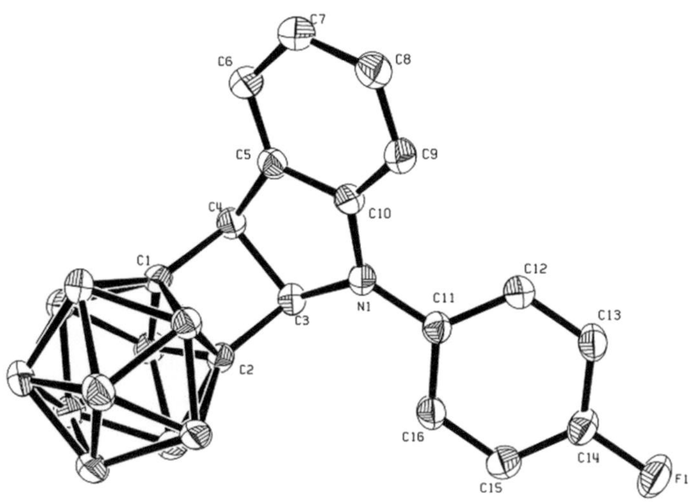

Fig. 3.9 Molecular structure of **III-4j**

To determine the proton source of the formal C–H insertion reaction, a deuterium-labeled substrate **III-3ea-d$_1$** was subjected to the reaction (Scheme 3.6). Product **III-5ea-d$_1$** with 84 % D incorporation at the cage C–H was obtained in 78 % yield (Figs. 3.14 and 3.15). No deuterium-labeled product was observed

Table 3.4 Substrate scope for regioselective carboranylation of *N*-methyl indoles[a]

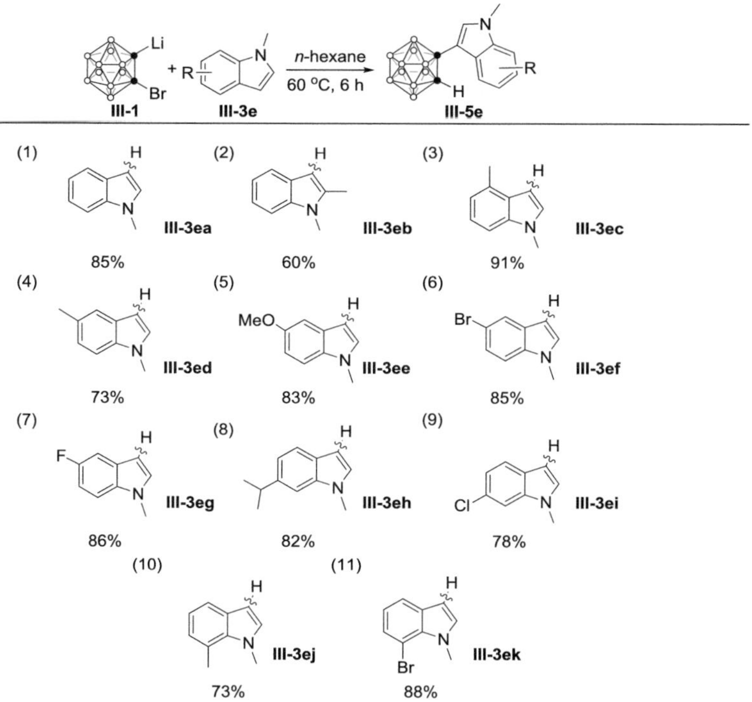

[a]Reaction conditions: **III-1** (1.0 mmol), indole **III-3e** (1.1 mmol) in *n*-hexane (10 mL), 60 °C, 6 h. Yields of isolated products were given

when the reaction was either carried out in cyclohexane-d_{12} or quenched with D_2O (Eqs. 3.7–3.8). These findings reveal that D (H) incorporated into the cage C–H only comes from the indole substrate. However, this deuterium was labile toward air and moisture as monitor by ^1H and ^2H NMR spectra (Figs. 3.15, 3.16, 3.17, 3.18, 3.19 and 3.20).

On the basis of the above experimental results, a plausible mechanism is proposed in Scheme 3.7. Elimination of LiBr from precursor **III-1** gives the reactive intermediate *o*-carboryne **III-2**. Electrophilic attack at the C3 position of *N*-protected indole **III-3** generates a zwitterionic intermediate **III-A**. **III-A** then undergoes either a proton shift to afford the formal insertion product **III-B** through pathway a or an intramolecular nucleophilic addition to yield the [2 + 2] cycloaddition product **III-C** through pathway b. Pathway b is reversible as evidenced by the thermal transformation of **III-4j** to **III-5j** (Scheme 3.3). The chemoselectivity of the reaction is dependent upon the nature of protecting group R. When R = alkyl groups, **III-B** is the sole product. When R = TMS group, **III-C** is the sole product. When R = aryl groups, both **III-B** and **III-C** are formed.

Table 3.5 Substrate scope for regioselective carboranylation of heterocycles[a]

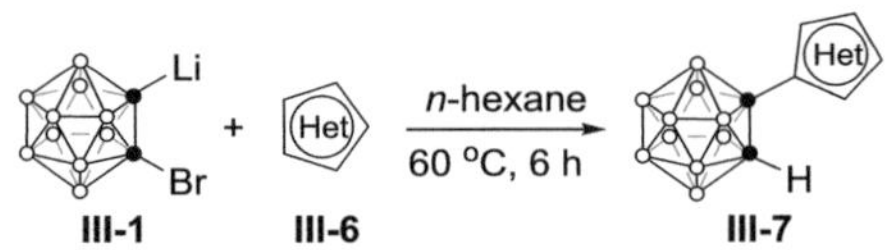

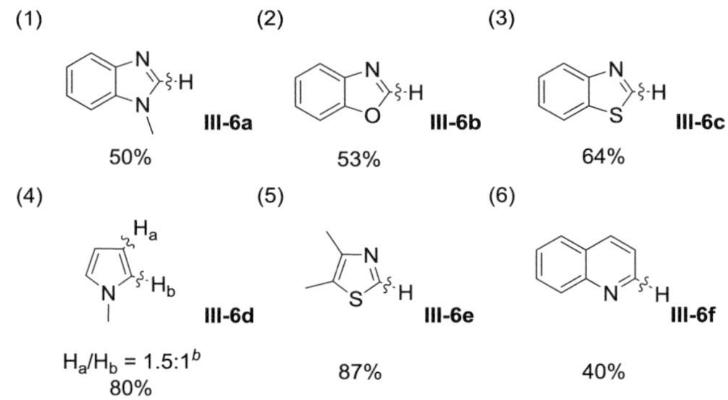

[a]Reaction conditions: **III-1** (1.0 mmol), *N*-heterocycle **III-6** (1.1 mmol) in *n*-hexane (10 mL), 60 °C, 6 h. Yields of isolated products were given

[b]Determined by ^{1}H NMR

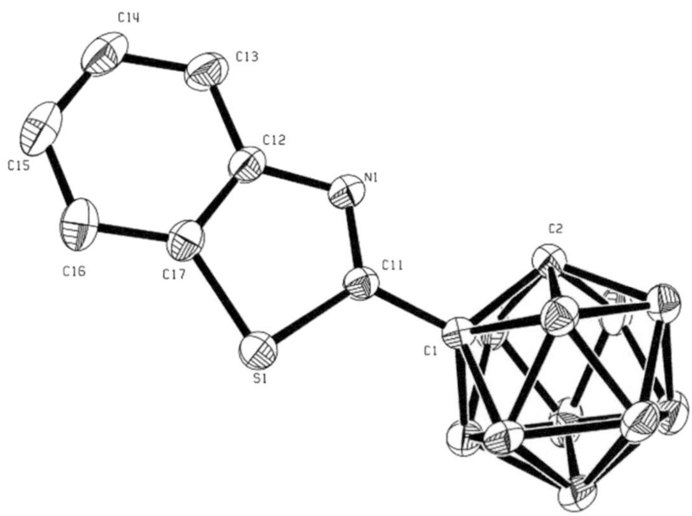

Fig. 3.10 Molecular structure of **III-7c**

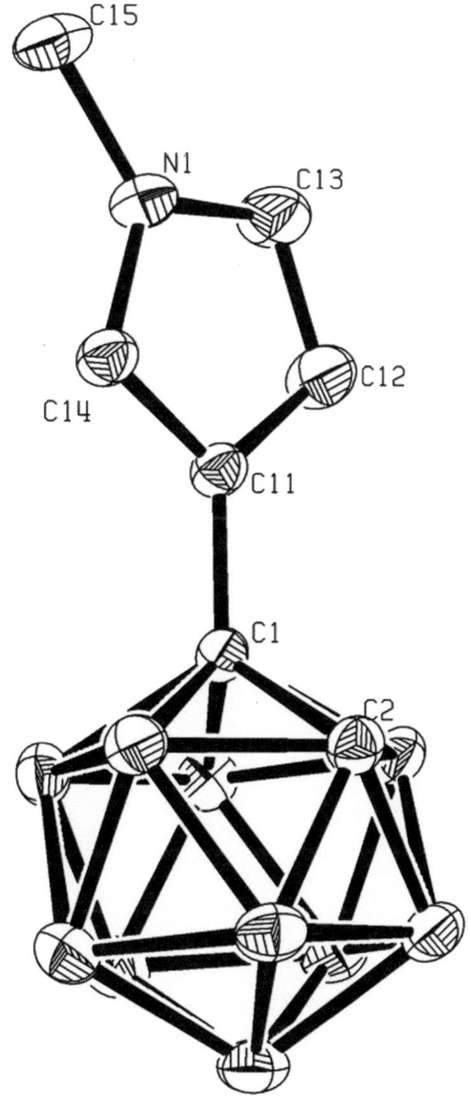

Fig. 3.11 Molecular structure of **III-7da**

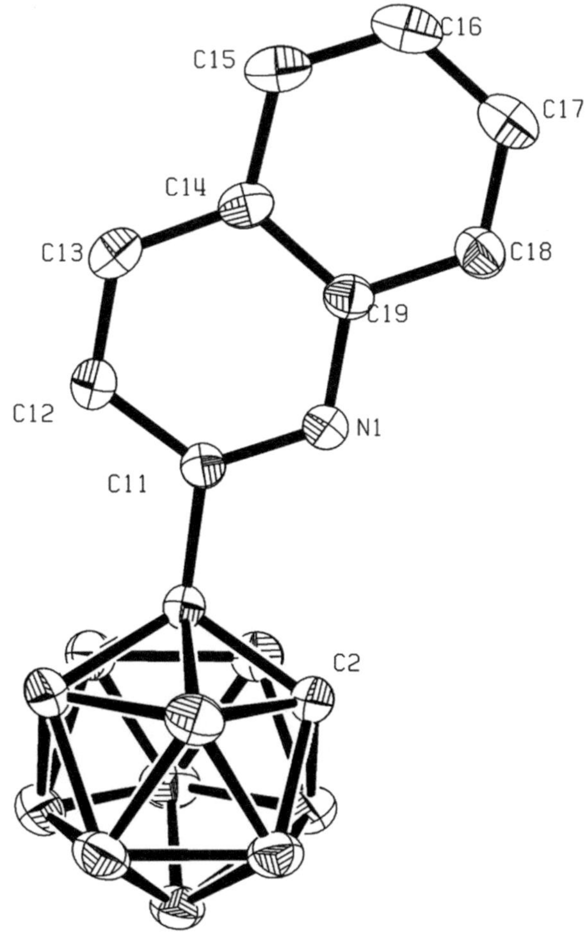

Fig. 3.12 Molecular structure of **III-7f**

Such **III-C/III-B** ratios increase if an electron-withdrawing substituent is introduced to the phenyl ring, which may be ascribed to the increased electrophilicity of C2 position in the intermediate **III-A** [25].

3.8 Summary

This work demonstrates that *o*-carboryne can serve as useful synthon for the synthesis of carborane-containing heterocycles. Reaction of *o*-carboryne with *N*-protected indoles goes two ways: dearomative [2 + 2] cycloaddition and formal C–H

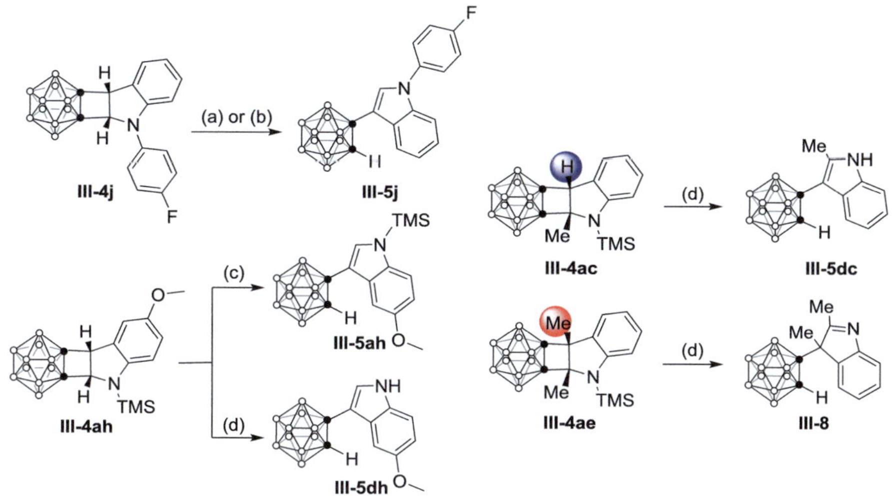

Scheme 3.3 Transformation of [2 + 2] cycloadducts[a]

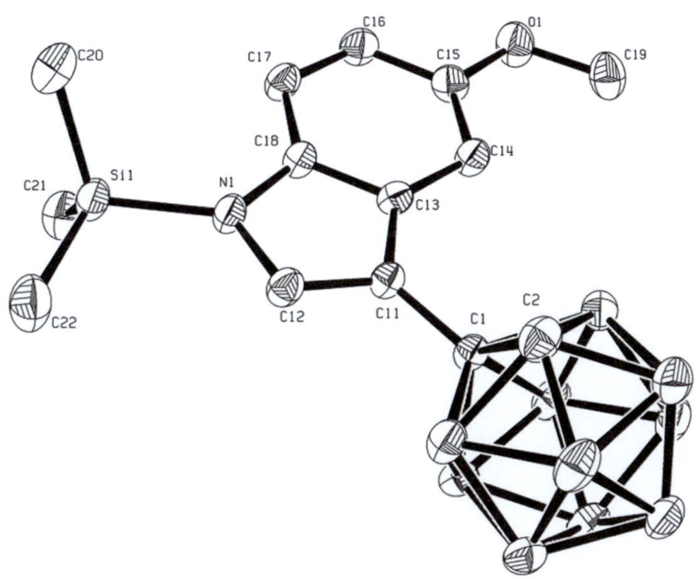

Fig. 3.13 Molecular structure of **III-5ah**

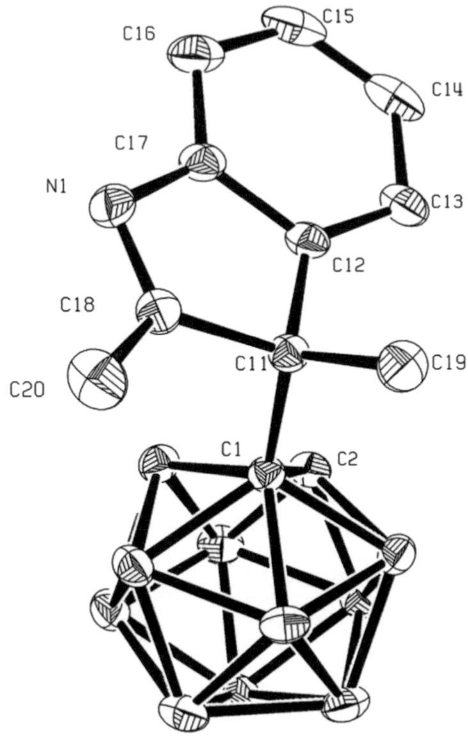

Fig. 3.14 Molecular structure of **III-8**

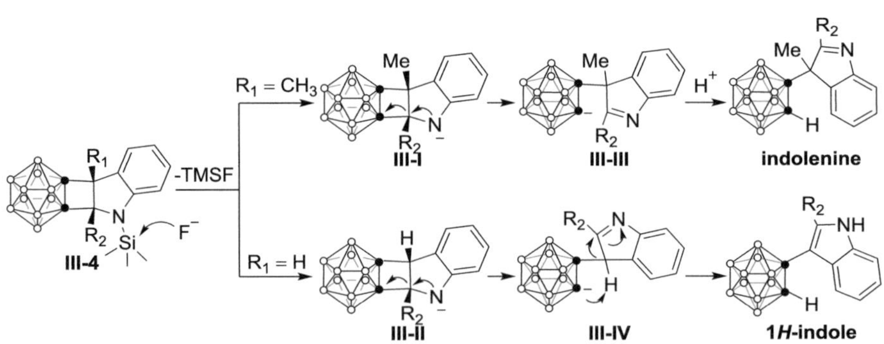

Scheme 3.4 Proposed reaction pathways for desilylation of [2 + 2] cycloadducts

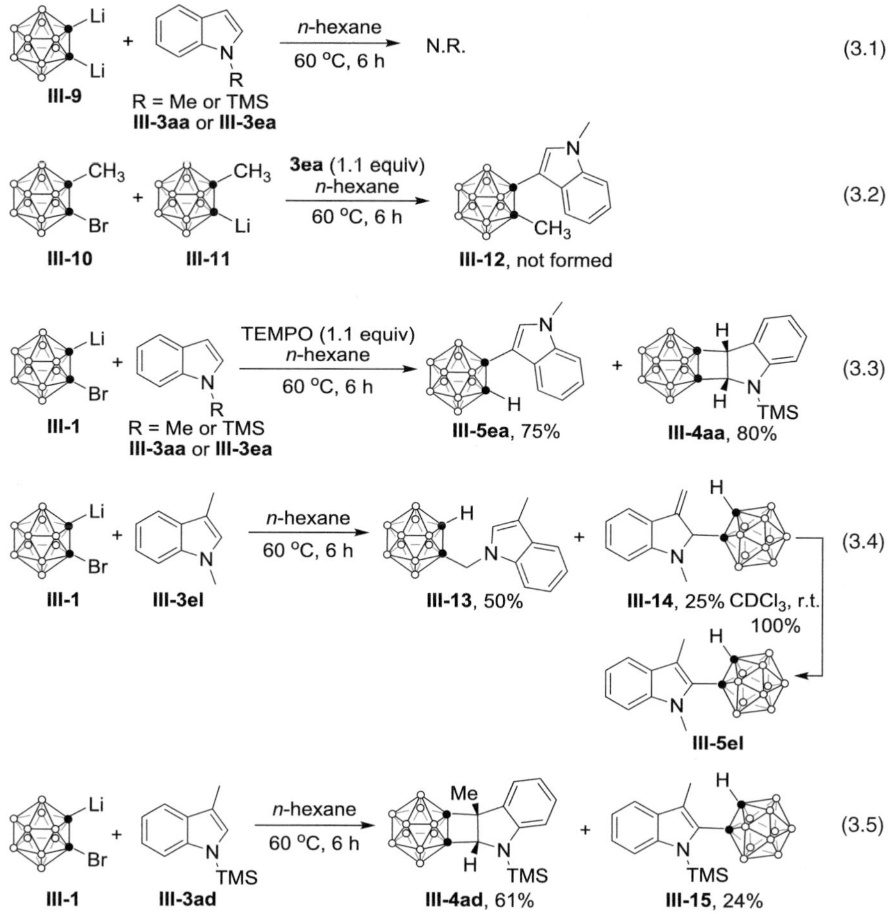

Scheme 3.5 Control experiments

insertion reaction, dependent upon the nature of *N*-substituents, leading to the synthesis of a series of carborane-functionalized indoles and indolines in a highly controlled manner. For *N*-TMS indoles, only dearomative [2 + 2] cycloaddition products are isolated, while the formal C–H insertion species are the sole products for *N*-alkyl ones. On the other hand, both [2 + 2] cycloadducts and formal C–H insertion products are isolated if *N*-substituents are aryls, in which the molar ratios of two products are determined by electronic properties of substituents on phenyl

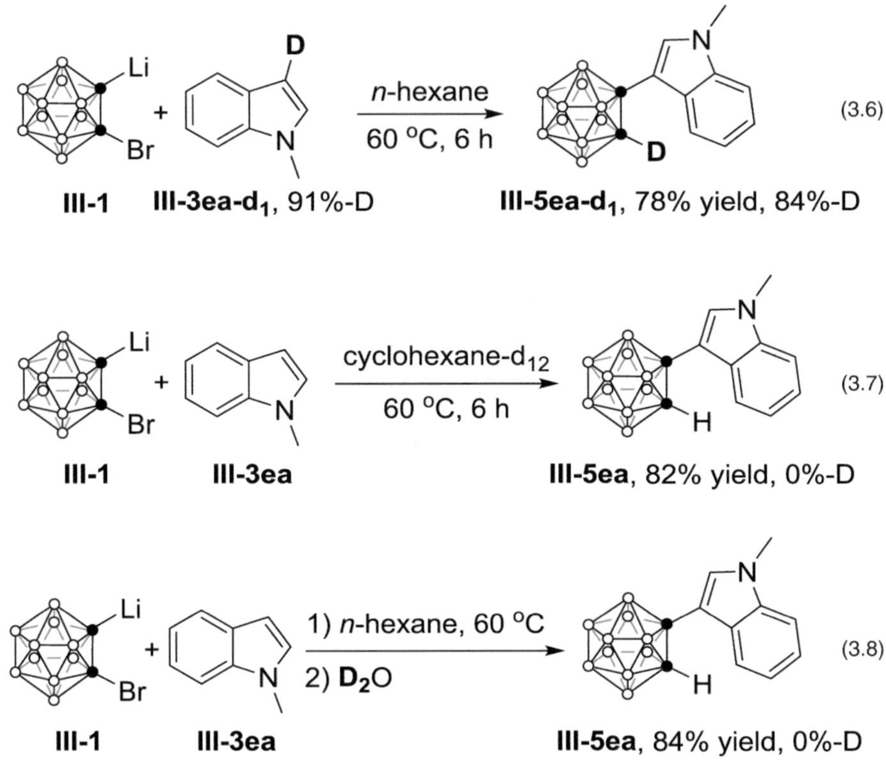

Scheme 3.6 Deuterium experiments

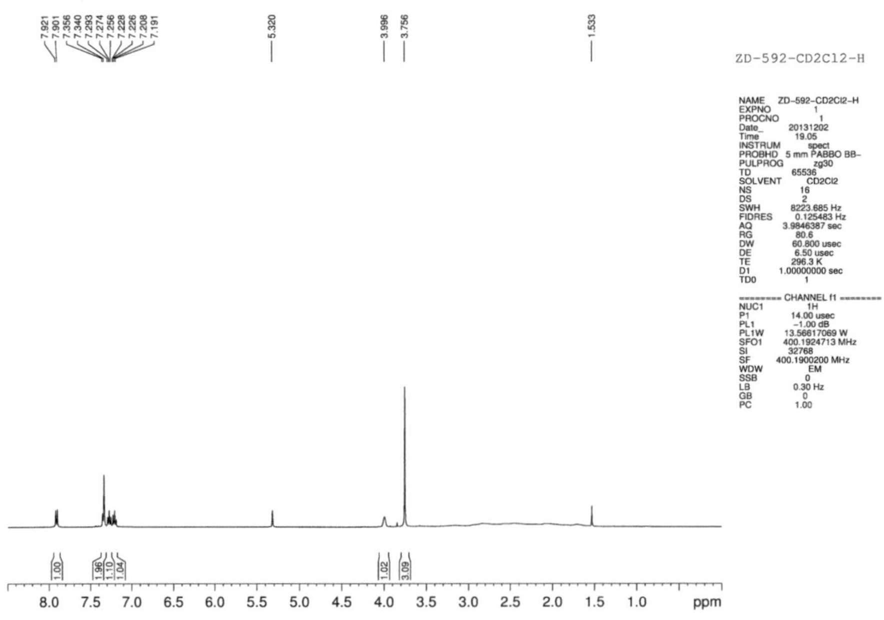

Fig. 3.15 ^{1}H NMR spectrum of **III-5ea** (for reference)

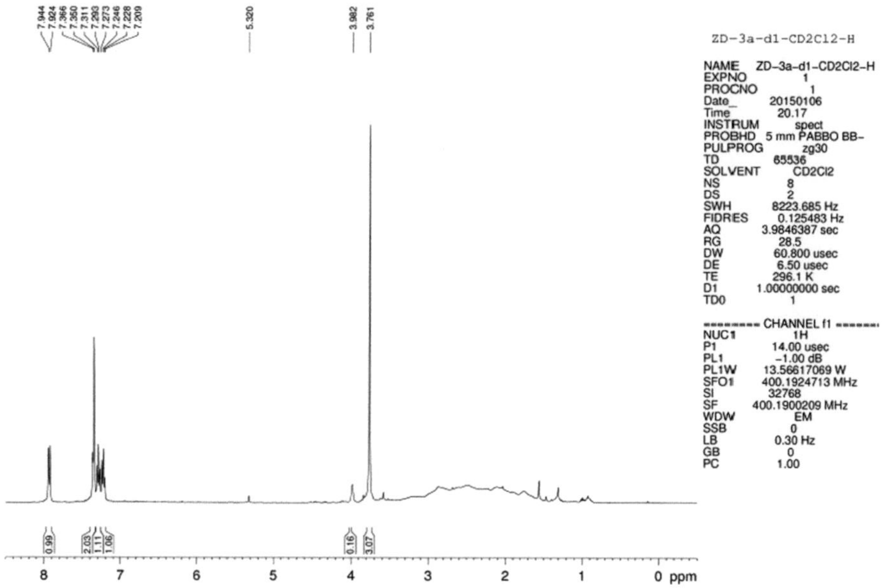

Fig. 3.16 ^{1}H NMR spectrum of **III-5ea–d$_1$** (recrystallized from the reaction mixture), 84 %-D

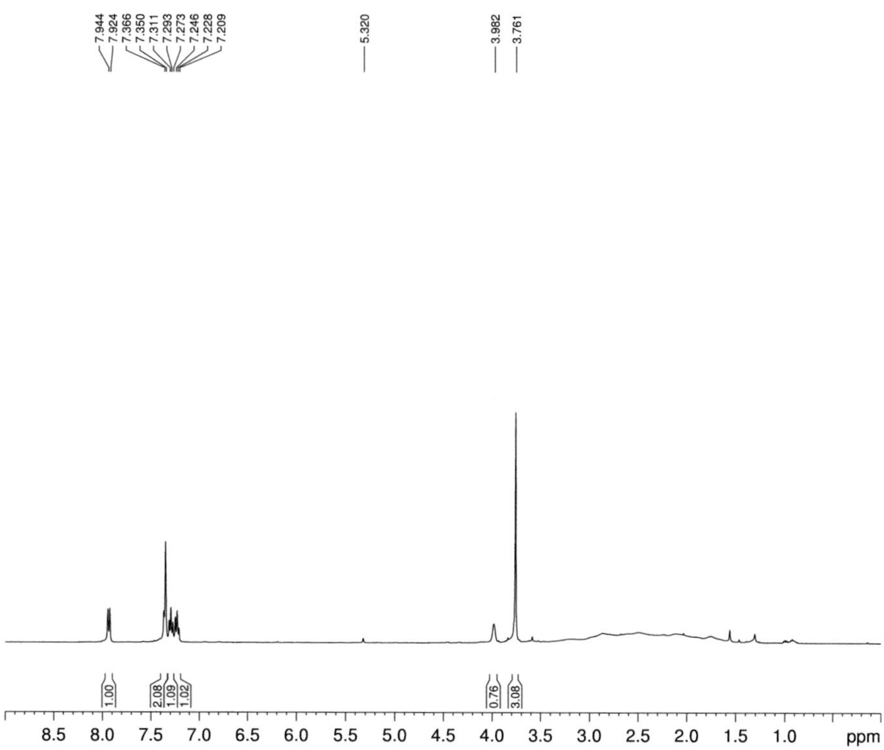

Fig. 3.17 ^{1}H NMR spectrum of the **III-5ea–d$_1$** (8 h at room temperature in $CDCl_3$) 24 %-D

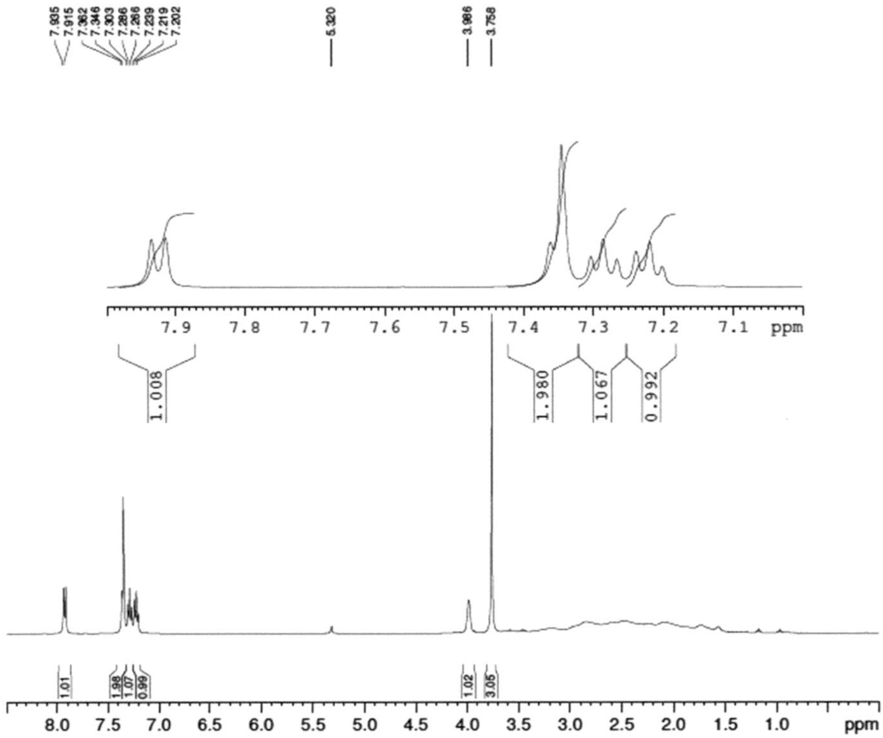

Fig. 3.18 ^{1}H NMR spectrum of **III-5ea–d$_1$** (after 3 days at room temperature), 0 %-D

ring. In general, electronic-donating substituents on *N* atom favor the formal insertion reaction, whereas electronic-withdrawing groups promote the dearomative [2 + 2] cycloaddition. To account for these experimental observations, a stepwise reaction mechanism is proposed.

It has been documented that indole molecules can undergo [2 + 2] cycloaddition reaction under UV irradiation or in the presence of transition metals. However, there is no precedent for indoles to undergo [2 + 2] cycloaddition without UV irradiation or transition metals. Thus, this work represents the first example observed for indoles to have such [2 + 2] cycloaddition in the absence of transition metals or without light. The resultant carborane-functionalized heterocycles may find their applications in medicines and materials science.

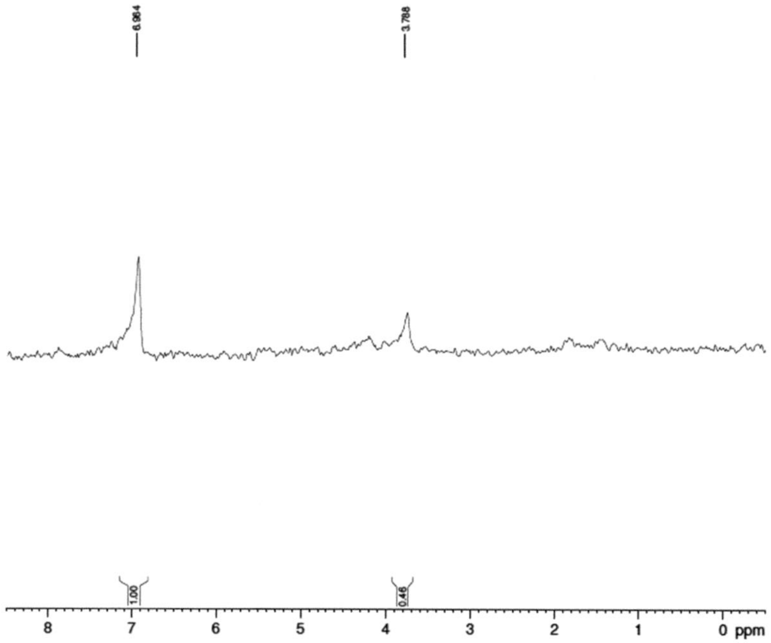

Fig. 3.19 ²H NMR spectrum of reaction mixture (*n*-hexane, reaction time 2 h), (**III-3ea–d$_1$**, δ 6.96; **III-5ea–d$_1$**, δ 3.79)

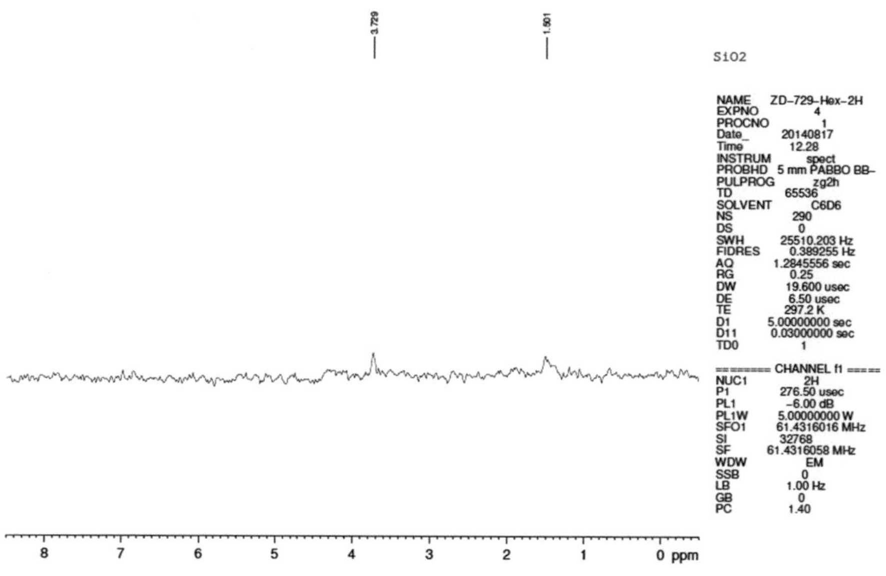

Fig. 3.20 ²H NMR spectrum of the above reaction mixture in the presence of H_2O and silica gel, ²H signal of **III-3ea–d$_1$** and **III-5ea–d$_1$** disappeared (**III-5ea–d$_1$**, δ 3.73; D_2O, δ 1.50)

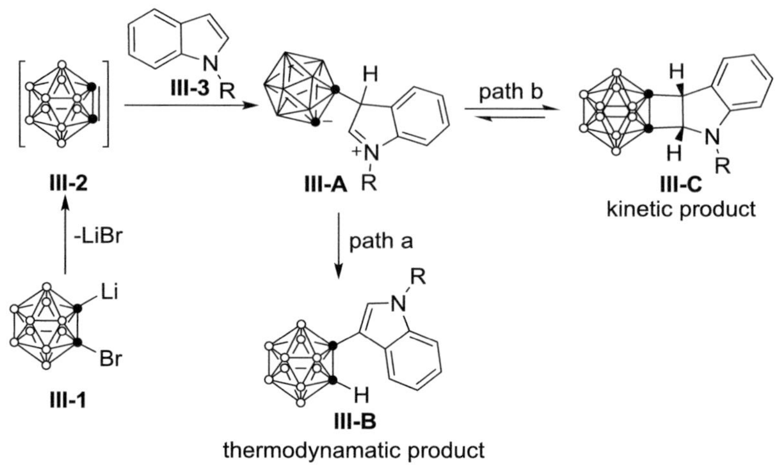

Scheme 3.7 Proposed reaction pathways

References

1. (a) Pouysegu L, Deffieux D, Quideau S (2010) Hypervalent iodine-mediated phenol dearomatization in natural product synthesis. Tetrahedron 66(13):2235–2261 (b) Roche SP, Porco JA (2011) Dearomatization strategies in the synthesis of complex natural products. Angew Chem Int Ed 50(18):4068–4093 (c) Zhuo C, Zhang W, You SL (2012) Catalytic asymmetric dearomatization reactions. Angew Chem Int Ed 51(51):12662–12686 (d) Ding QP, Zhou XL, Fan RH (2014) Recent advances in dearomatization of heteroaromatic compounds. Org Biomol Chem 12(27):4807–4815
2. Roche SP, Tendoung JJY, Treguier B (2015) Advances in dearomatization strategies of indoles. Tetrahedron 71(22):3549–3591
3. (a) Zhang D, Song H, Qin Y (2011) Total synthesis of indoline alkaloids: a cyclopropanation strategy. Acc Chem Res 44(6):447–457 (b) Wang HX, Reisman SE (2014) Enantioselective total synthesis of (-)- lansai B and (+)- nocardioazines A and B. Angew Chem Int Ed 53 (24):6206–6210 (c) Spangler JE, Davies HML (2013) Catalytic asymmetric synthesis of pyrroloindolines via a rhodium(II)-catalyzed annulation of indoles. J Am Chem Soc 135 (18):6802–6805 (d) Repka LM, Ni J, Reisman SE (2010) Enantioselective synthesis of pyrroloindolines by a formal [3 + 2] cycloaddition reaction. J Am Chem Soc 132(41):14418–14420 (e) Li H, Hughes RP, Wu J (2014) Dearomative indole (3 + 2) cycloaddition reactions. J Am Chem Soc 136(17):6288–6296 (f) Kawano M, Kiuchi T, Negishi S et al (2013) Regioselective inter- and intramolecular formal [4 + 2] cycloaddition of cyclobutanones with indoles and total synthesis of (±)-aspidospermidine. Angew Chem Int Ed 52(3):906–910 (g) Zhang GZ, Huang XG, Li GT et al (2008) Au-containing all-carbon 1,4-dipoles: Generation and [4 + 2] annulation in the formation of carbo-/heterocycles. J Am Chem Soc 130(6):1814–1815. (h) Martin DBC, Vanderwal CD (2009) Efficient access to the core of the strychnos, aspidosperma and Iboga alkaloids. A short synthesis of norfluorocurarine. J Am Chem Soc 131(10):3472–3473 (i) Robertson FJ, Kenimer BD, Wu J (2011) Direct annulation and alkylation of indoles with 2-aminobenzyl alcohols catalyzed by TFA. Tetrahedron 67 (24):4327–4332
4. (a) England DB, Kuss TDO, Keddy RG et al (2001) Cyclopentannulation of 3-alkylindoles: a synthesis of a tetracyclic subunit of the kopsane alkaloids. J Org Chem 66(13):4704–4709

(b) Bajtos B, Yu M, Zhao HD et al (2007) C-2/C-3 annulation and C-2 alkylation of indoles with 2-alkoxycyclopropanoate esters. J Am Chem Soc 129(31):9631–9634 (c) Barluenga J, Tudela E, Ballesteros A et al (2009) Asymmetric C2-C3 cyclopentannulation of the indole ring. J Am Chem Soc 131(6):2096–2097 (d) Lian YJ, Davies HML (2010) Rhodium-catalyzed [3 + 2] annulation of indoles. J Am Chem Soc 132(2):440–441 (e) Xiong H, Xu H, Liao SH et al (2013) Copper-catalyzed highly enantioselective cyclopentannulation of indoles with donor-acceptor cyclopropanes. J Am Chem Soc 135(21):7851–7854

5. (a) Wenkert E, Moeller PDR, Piettre SR (1988) 5-membered aromatic heterocycles as dienophiles in diels-alder reactions—furan, pyrrole, and indole. J Am Chem Soc 110 (21):7188–7194 (b) Hsieh MF, Rao PD, Liao CC (1999) Diels-alder and michael addition reactions of indoles with masked o-benzoquinones: synthesis of highly functionalized hydrocarbazoles and 3-arylindoles. Chem Comm (15):1441–1442 (c) Biolatto B, Kneeteman M, Paredes E et al (2001) Reactions of 1-tosyl-3-substituted indoles with conjugated dienes under thermal and/or high-pressure conditions. J Org Chem 66(11):3906–3912 (d) Cai Q, You SL (2012) Organocatalyzed enantioselective formal [4 + 2] cycloadclition of 2,3-disubstituted indole and methyl vinyl ketone. Org Lett 14(12):3040–3043
6. Mei GJ, Yuan H, Gu YQ et al (2014) Dearomative indole [5 + 2] cycloaddition reactions: Stereoselective synthesis of highly functionalized cyclohepta[b]indoles. Angew Chem Int Ed 53(41):11051–11055
7. Gribble GW (2010) Heterocyclic scaffolds II: topics in heterocyclic chemistry. Springer, Berlin
8. (a) Zhang LM (2005) Tandem au-catalyzed 3,3-rearrangement-[2 + 2] cycloadditions of propargylic esters: Expeditious access to highly functionalized 2,3-indoline-fused cyclobutanes. J Am Chem Soc 127(48):16804–16805 (b) Jia MQ, Monari M, Yang QQ et al (2015) Enantioselective gold catalyzed dearomative [2 + 2]-cycloaddition between indoles and allenamides. Chem Comm 51(12):2320–2323 (c) Faustino H, Bernal P, Castedo L et al (2012) Gold(i)-catalyzed intermolecular [2 + 2]cycloadditions between allenamides and alkenes. Adv Synth Catal 354(9):1658–1664
9. (a) Pozharskii AF, Soldatenkov AT, Katrizky AR (2011) Heterocycles in life and society: an introduction to heterocyclic chemistry, biochemistry and applications, 2nd edn. Wiley, Chichester (b) Quin LD, Tyrell JA (2010) Fundamentals of heterocyclic chemistry: importance in nature and in the synthesis of pharmaceuticals. Wiley, New York
10. (a) Sundberg RJ (1970) The chemistry of indoles. Academic Press, New York (b) Houlihan WJ. Indoles. Wiley-Interscience, New York (c) Thomas EJ (2000) Science of synthesis. Thieme, Stuttgart (d) Katritzky AR, Scriven EFV, Rees CW (2008) Comprehensive heterocyclic chemistry III. Elsevier, Oxford
11. (a) Cacchi S, Fabrizi G (2005) Synthesis and functionalization of indoles through palladium-catalyzed reactions. Chem Rev 105(7):2873–2920 (b) Beccalli EM, Broggini G, Martinelli M et al (2007) C–C, C–O, C–N bond formation on sp^2 carbon by Pd(II)-catalyzed reactions involving oxidant agents. Chem Rev 107(11):5318–5365 (c) Seregin IV, Gevorgyan V (2007) Direct transition metal-catalyzed functionalization of heteroaromatic compounds. Chem Soc Rev 36(7):1173–1193 (d) Bandini M, Eichholzer A (2009) Catalytic functionalization of indoles in a new dimension. Angew Chem Int Ed 48(51):9608–9644 (e) Lyons TW, Sanford MS (2010) Palladium-catalyzed ligand-directed C–H functionalization reactions. Chem Rev 110(2):1147–1169
12. a) Bandini M, Melloni A, Umani-Ronchi A (2004) New catalytic approaches in the stereoselective friedel-crafts alkylation reaction. Angew Chem Int Ed 43 (5):550–556 (b) You SL, Cai Q, Zeng M (2009) Chiral bronsted acid catalyzed friedel-crafts alkylation reactions. Chem Soc Rev 38 (8):2190-2201 (c) Poulsen TB, Jorgensen KA (2008) Catalytic asymmetric friedel-crafts alkylation reactions-copper showed the way. Chem Rev 108 (8):2903-2915
13. (a) Kracke GR, VanGordon MR, Sevryugina YV et al (2015) Carborane-derived local anesthetics are isomer dependent. Chemmedchem 10(1):62–67 (b) Wilkinson SM,

Gunosewoyo H, Barron ML et al (2014) The first cns-active carborane: a novel P2X7 receptor antagonist with antidepressant activity. Acs Chem Neurosci 5(5):335–339 (c) Scholz M, Kaluderovic GN, Kommera H et al (2011) Carbaboranes as pharmacophores: similarities and differences between aspirin and asborin. Eur J Med Chem 46(4):1131–1139 (d) Beer ML, Lemon J, Valliant JF (2010) Preparation and evaluation of carborane analogues of tamoxifen. J Med Chem 53(22):8012–8020 (e) Julius RL, Farha OK, Chiang J et al (2007) Synthesis and evaluation of transthyretin amyloidosis inhibitors containing carborane pharmacophores. Proc Natl Acad Sci USA 104(12):4808–4813

14. (a) Visbal R, Ospino I, Lopez-de-Luzuriaga JM et al (2013) N-heterocyclic carbene ligands as modulators of luminescence in three-coordinate gold(I) complexes with spectacular quantum yields. J Am Chem Soc 135(12):4712–4715 (b) Cioran AM, Musteti AD, Teixidor F et al (2012) Mercaptocarborane-capped gold nanoparticles: electron pools and ion traps with switchable hydrophilicity. J Am Chem Soc 134(1):212–221 (c) Koshino M, Tanaka T, Solin N et al (2007) Imaging of single organic molecules in motion. Science 316(5826):853–853 (d) Yinghuai Z, Peng AT, Carpenter K et al (2005) Substituted carborane-appended water-soluble single-wall carbon nanotubes: New approach to boron neutron capture therapy drug delivery. J Am Chem Soc 127(27):9875–9880
15. (a) Hosmane NS, Maguire JA (2007) d- and f-Block metallacarbaboranes. In: Crabtree RH, Mingos DMP (eds) Comprehensive organometallic chemistry III, Elsevier, Oxford (b) Xie Z (2002) Advances in the chemistry of metallacarboranes of f-block elements. Coord Chem Rev 231(1–2):23–46 (c) Xie Z (2003) Cyclopentadienyl-carboranyl hybrid compounds: a new class of versatile ligands for organometallic chemistry. 36(1):1–9 (d) Deng L, Xie Z (2007) Advances in the chemistry of carboranes and metallacarboranes with more than 12 vertices. Coord Chem Rev 251(17–20):2452–2476 (e) Yao ZJ, Jin GX (2013) Transition metal complexes based on carboranyl ligands containing N, P, and S donors: Synthesis, reactivity and applications. Coordin Chem Rev 257 (17–18):2522–2535 (f) Zhang J, Xie Z (2014) Synthesis, structure, and reactivity of 13-and 14-vertex carboranes. Acc Chem Res 47 (5):1623–1633
16. (a) Wee KR, Han WS, Cho DW et al (2012) Carborane photochemistry triggered by aryl substitution: carborane-based dyads with phenyl carbazoles. Angew Chem Int Ed 51 (11):2677–2680 (b) Wee KR, Cho YJ, Song JK et al (2013) Multiple photoluminescence from 1,2-dinaphthyl-ortho-carborane. Angew Chem Int Ed 52(37):9682–9685 (c) Wee KR, Cho YJ, Jeong S et al (2012) Carborane-based optoelectronically active organic molecules: wide band gap host materials for blue phosphorescence. J Am Chem Soc 134(43):17982–17990
17. Tietze LF, Griesbach U, Bothe U et al (2002) Novel carboranes with a DNA binding unit for the treatment of cancer by boron neutron capture therapy. Chem Bio Chem 3(2–3):219–225
18. Grimes RN (2011) Carboranes, 2nd edn. Academic Press, Amsterdam
19. (a) Hosmane NS (2011) Boron science: new technologies and applications. CRC Press, Boca Raton (b) Grimes RN (2015) Carboranes in the chemist's toolbox. Dalton Trans 44(13):5939–5956 (c) Bregadze VI (1992) Dicarba-closo-dodecaboranes $C_2B_{10}H_{12}$ and their derivatives. Chem Rev 92(2):209–223
20. (a) Qiu Z, Wang SR, Xie Z (2010) Nickel-catalyzed regioselective [2 + 2 + 2] cycloaddition of carboryne with alkynes. Angew Chem Int Edit 49(27):4649–4652 (b) Qiu Z, Xie Z (2014) Generation and reactivity of o-carborynes. Dalton Trans 43(13):4925–4934
21. Huang QR, Gingrich HL, Jones M (1991) Ene reactions of 1,2-dehydro-ortho-carborane. Inorg Chem 30(17):3254–3257
22. Wang SR, Qiu Z, Xie Z (2011) Regioselective insertion of carborynes into ethereal C-H bond: Facile synthesis of alpha-carboranylated ethers. J Am Chem Soc 133(15):5760–5763
23. Wang SR, Xie Z (2012) Formal insertion of o-carborynes into ferrocenyl C-H bonds: A simple access to o-carboranylferrocenes. Organometallics 31(12):4544–4550
24. Gona KB, Thota JLVNP, Baz Z et al (2015) Synthesis and C-11-radiolabelling of 2-carboranyl benzothiazoles. Molecules 20(5):7495–7508
25. Liu Q, Zhao QY, Liu J et al (2012) A trans diacyloxylation of indoles. Chem Commun 48 (26):3239–3241

Chapter 4
Reaction of *o*-Carboryne with Nitrones: A Formal [5 + 2] Cycloaddition

Abstract This chapter details the first example of transition-metal-free [5 + 2] cycloaddition of nitrones with *o*-carboryne. In this process, the nitrone moieties served as five-atom building blocks with both heteroatoms being incorporated into the framework of the final products. Mechanistic studies suggest that the N–O bond cleavage in the initially formed [3 + 2] cycloadducts, followed by an intramolecular nucleophilic addition, and rearomatization leads to the final products. This approach allows rapid construction of *o*-carborane-fused seven-membered heterocycles with a broad substrate scope, which also provides new insight into the chemistry of nitrones.

4.1 Introduction

Heterocyclic skeletons are widespread among natural products as well as bioactive molecules. As a result, developing synthetic methods for the efficient construction of heterocyclic compounds is now one of the most important parts of modern organic synthesis. Due to its advantage of multiple bond formation in a single step, cycloaddition reaction has created new opportunities to access heterocycles of great complexity [1].

Nitrones have been known for more than a century as electrophiles toward organometallic reagents or as versatile 1,3-dipolarophiles (Scheme 4.1). Nowadays, cycloadditions involving nitrones have been employed profusely in synthetic chemistry, which provide an easy access to five- and six-membered nitrogen heterocyclic frameworks that are found in many natural products [2]. It has been well documented that nitrones can readily undergo [3 + 2] [3], [3 + 3] [3–8], [4 + 3] [9, 10], and [2 + 2 + 3] [11] cycloadditions (Fig. 4.1). In these intriguing studies, nitrones serve exclusively as three-atom building blocks to afford five-, six-, and seven-membered heterocycles. In 2014, Wan and coworkers reported a rhodium-catalyzed cyclization of diynes with nitrones in which the nitrone moiety was employed as five-atom coupling partner in the cyclization process (Scheme 4.3) [12].

D. Zhao, *Functionalization of Carborane via Carboryne Intermediates*,
Springer Theses, DOI 10.1007/978-981-10-1569-4_4

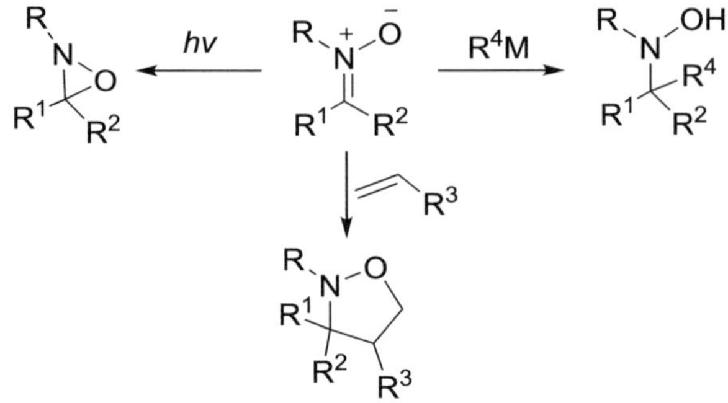

Scheme 4.1 Common reactivity patterns of nitrones

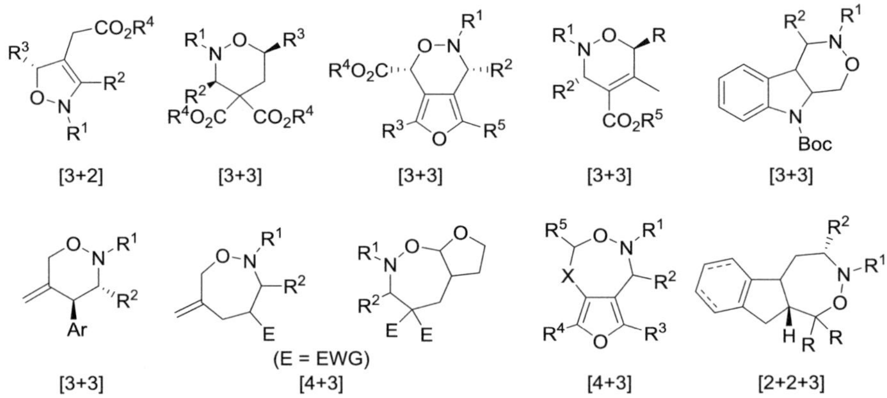

Fig. 4.1 Cycloaddition products of nitrones

On the other hand, 1,2-dehydro-*o*-carborane (*o*-carboryne) (**IV-2**), which can be viewed as a three-dimensional relative of benzyne, can undergo cycloadditions, ene reaction, and C–H bond insertion reaction with a variety of organic molecules to afford a large class of functional *o*-carboranes [13]. Cycloadditions involving *o*-carboryne intermediate have been developed to enable the synthesis of various carbocyclic carborane derivatives. However, they are only limited to [4 + 2] [13–19] and [2 + 2] [20–22] fashions. Moreover, cycloadditions of *o*-carboryne for the synthesis of carborane-fused heterocycles still remain unrevealed [13d].

4.2 Reaction of *o*-Carboryne with Nitrones

In Chaps. 2 and 3, we have extensively investigated the reactivity of *o*-carboryne derived from 1-X-2-lithio-*o*-carborane (X = I, Br) [13a, c], which are strong bases. As part of our continuous efforts in exploring the reactivity of *o*-carboryne, we draw

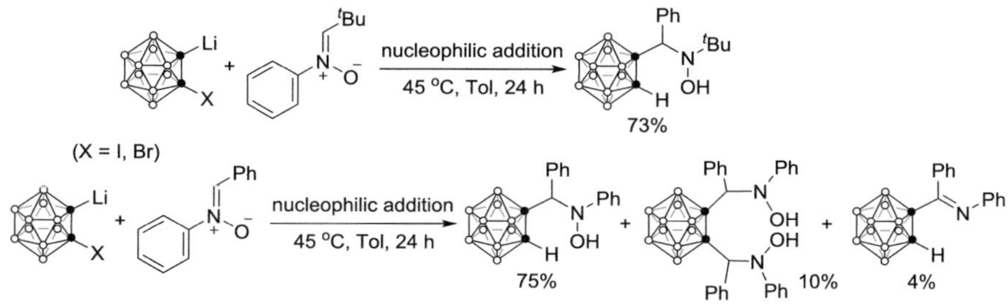

Scheme 4.2 Reaction of nitrones with 1-X-2-lithio-*o*-carborane (X = I, Br)

our attention to the reaction of nitrones with *o*-carboryne. Thus, the reaction of nitrones with the above two precursors was first examined. However, similar to the reaction of nitrone with other organometallic reagents, only nucleophilic addition reaction was observed (Scheme 4.2) [23].

We wonder whether this undesired nucleophilic addition reaction could be suppressed if a neutral precursor phenyl[*o*-(trimethylsilyl)carboranyl]iodonium acetate (**IV-1**), which was first reported by Kang and Ko [13b], was utilized as *o*-carboryne source. Surprisingly, an unexpected formal [5 + 2] cycloaddition takes place instead of the [3 + 2] cycloaddition reaction to afford carborane-fused seven-membered heterocycles in good yields. It is noted that benzisoxazolines are generated from the reaction of nitrones with arynes via [3 + 2] cycloadditions (Scheme 4.3) [24]. The experimental details and mechanistic studies are discussed in this section.

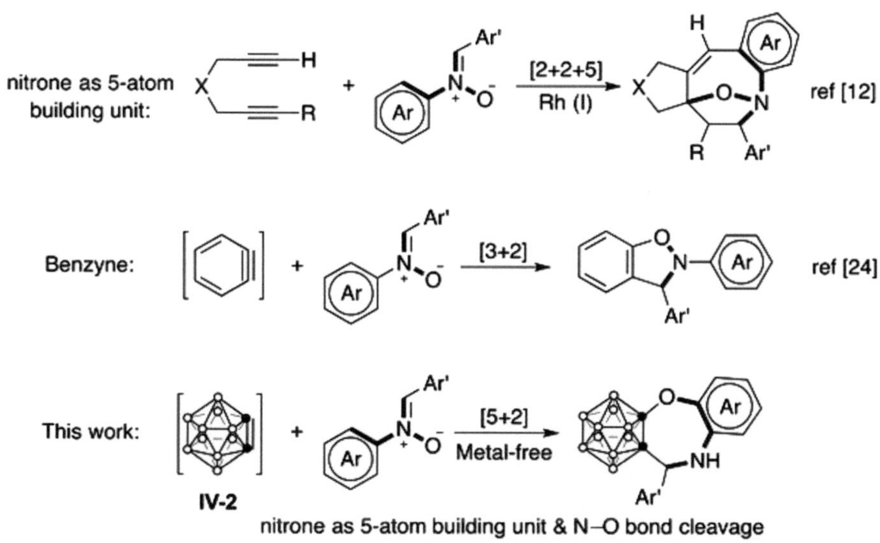

Scheme 4.3 Cycloadditions involving nitrones, benzyne, and *o*-Carboryne

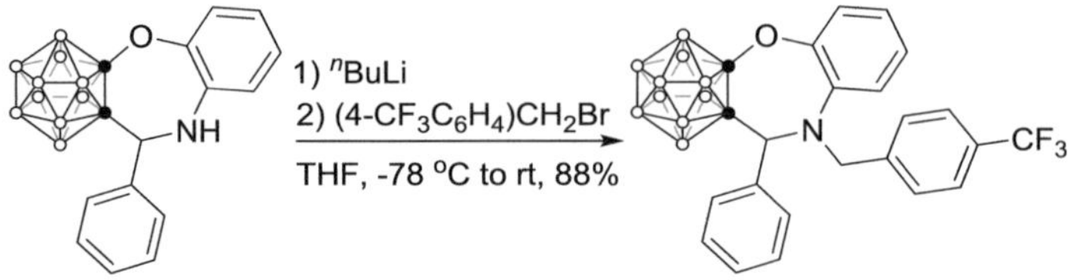

Scheme 4.4 Synthesis and structural determination of adduct **IV-5a**

In our initial investigation, a THF solution of phenyl[*o*-(trimethylsilyl)carboranyl]iodonium acetate (**IV-1**) [13b] was treated with *N*, alpha-diphenyl nitrone (**IV-3a**) in the presence of CsF. Surprisingly, the expected [3 + 2] cycloadduct was not observed in the ^{1}H NMR spectrum of the crude reaction mixture. Instead, an oily compound was isolated in 72 % yield as the sole product, and it was finally confirmed as a [5 + 2] cycloaddition product **IV-4a** (Scheme 4.4 and Fig. 4.2).

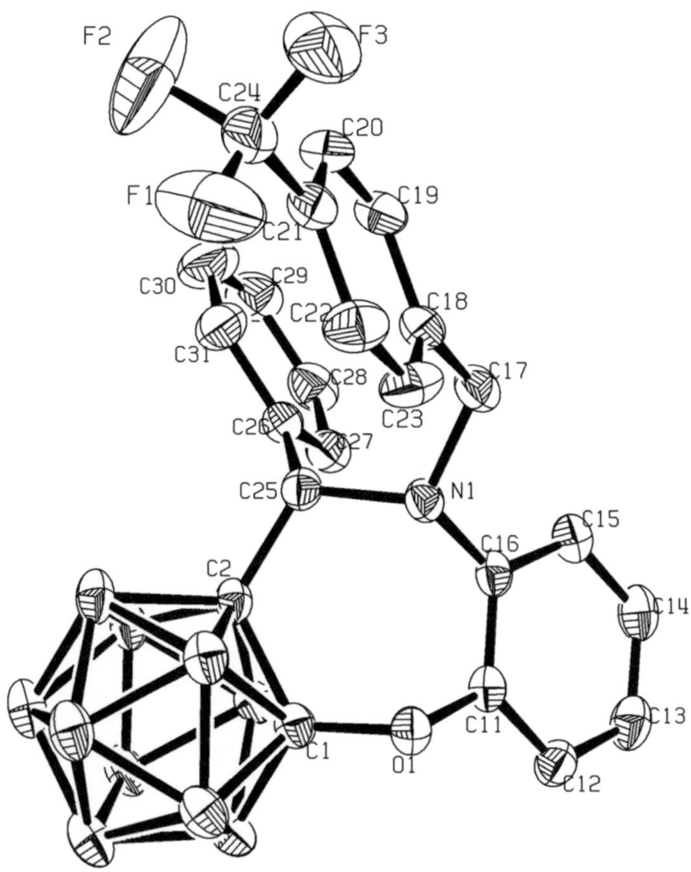

Fig. 4.2 Molecular structure of **IV-5a**

Further optimization of the reaction conditions found that both temperature and reaction time are crucial to the yields of the product. High temperatures or prolonged reaction time dramatically decreased the yields, probably due to the deboronation of the product by fluoride (Table 4.1, entries 1–4) [25]. The reaction medium also displayed significant influence on the outcome of this reaction. *o*-Carborane (*o*-$C_2B_{10}H_{12}$) was produced quantitatively when acetonitrile was employed as solvent, and a complex mixture was obtained if the reaction was performed in dichloromethane (Table 4.1, entries 5–6). Decreasing the amount of CsF to 1.0 or 1.5 equiv led to the incomplete conversion of precursor **IV-1** (Table 4.1, entries 7 and 8) [24e]. A lower yield was obtained when increasing the amount of nitrone **IV-3a** to 2.0 equiv (Table 4.1, entry 9).

Under the optimal conditions shown in Table 4.1, entry 1, the scope of this [5 + 2] cycloaddition reaction of *o*-carboryne with various nitrones was examined and the desired cycloadducts were isolated in moderate to good yields (Table 4.2). Both electron-withdrawing groups (Cl, Br, F, CF_3, and NO_2) and electron-donating groups (Me and OMe) on the C– and N–aryl rings were well tolerated. In general, nitrones with electron-donating substituents tend to result in higher yields than those with electron-withdrawing ones (**IV-3d–g** vs. **IV-3h–i**; **IV-3m–n** vs. **IV-3o–q**). For instance, reactions of nitrone **IV-3h** and **IV-3m** proceeded smoothly to afford the cycloadducts in good yields, whereas only 31 or 45 % of the desired product was isolated for nitrone **IV-3g** or **IV-3q** that bearing a nitro group, respectively.

Table 4.1 Optimization of reaction conditions[a]

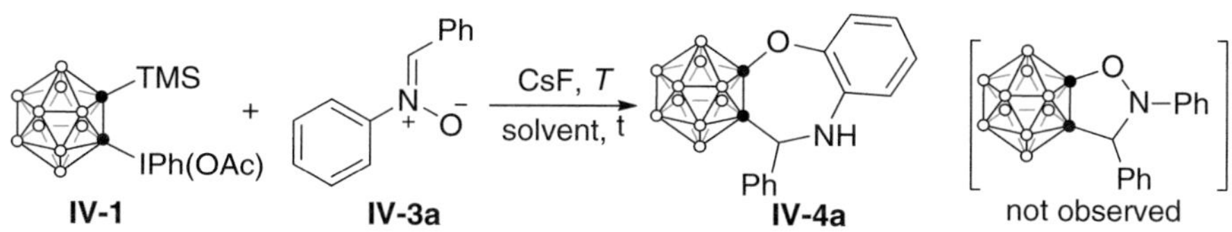

Entry	**IV-3a** (equiv)	CsF (equiv)	Solvent	T (°C)	t (min)	**IV-4a** (%)[b]
1	1.1	2.0	THF	25	30	78
2	1.1	2.0	THF	25	120	73
3	1.1	2.0	THF	60	120	50
4	1.1	2.0	THF	60	720	15
5	1.1	2.0	CH_3CN	25	30	–[c]
6	1.1	2.0	CH_2Cl_2	25	30	–[d]
7	1.1	1.0	THF	25	30	45[e]
8	1.1	1.5	THF	25	30	69[e]
9	2.0	2.0	THF	25	30	57

[a]Reaction conditions: **IV-1** (0.1 mmol), **IV-3a** (0.11 or 2.0 mmol), CsF (2.0 or 1.0 mmol), solvent (3 mL)
[b]Isolated yields
[c]Main product is *o*-carborane
[d]Complex mixture
[e]**IV-1** was not completely consumed

Table 4.2 [5 + 2] cycloaddition of nitrones with *o*-carboryne[a]

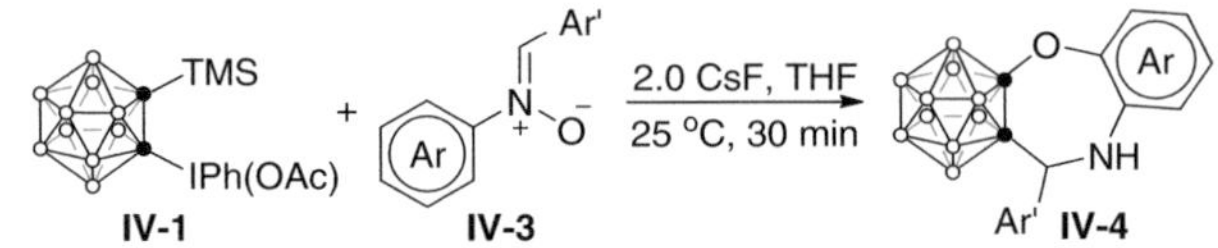

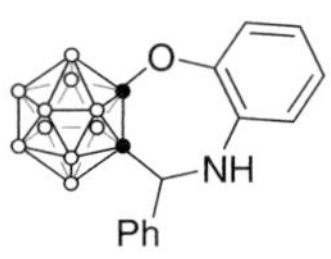

IV-4a, 78%

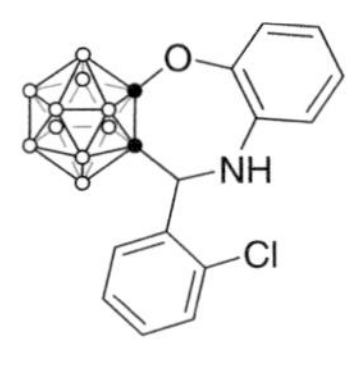

IV-4b, 67%

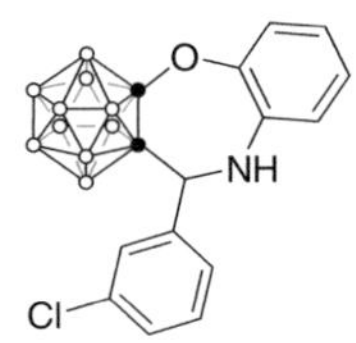

IV-4c, 76%

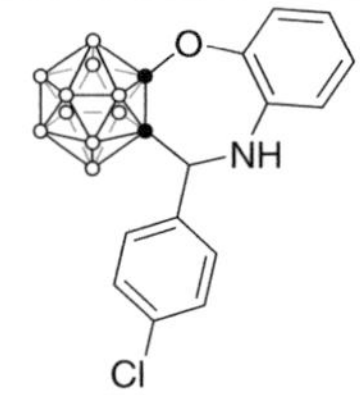

IV-4d, 72%

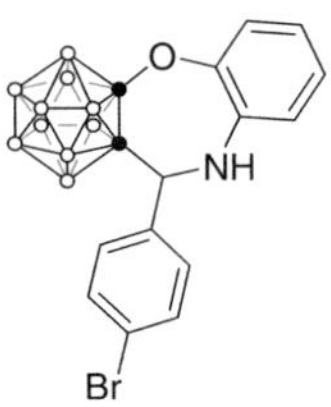

IV-4e, 75%

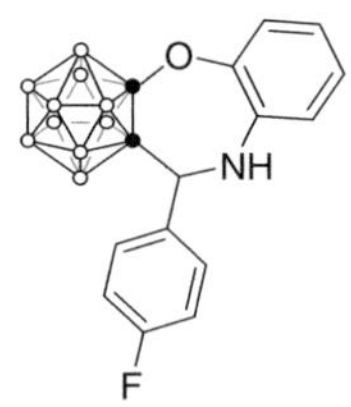

IV-4f, 68%

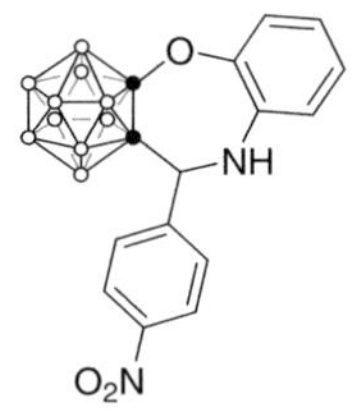

IV-4g, 31%

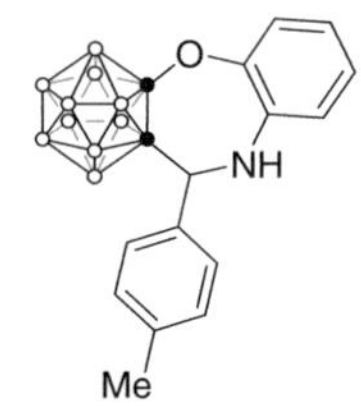

IV-4h, 82%

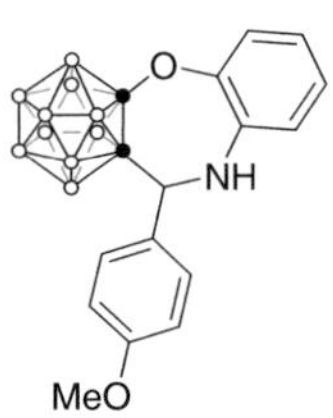

IV-4i, 77%

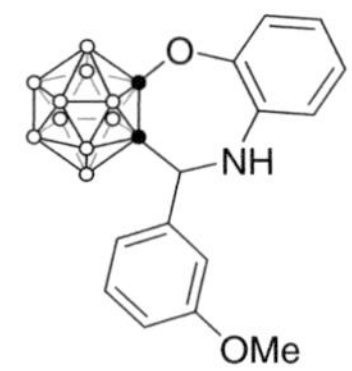

IV-4j, 62%

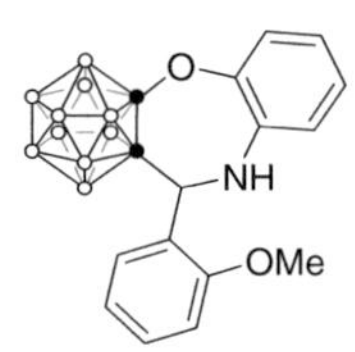

IV-4k, 71%

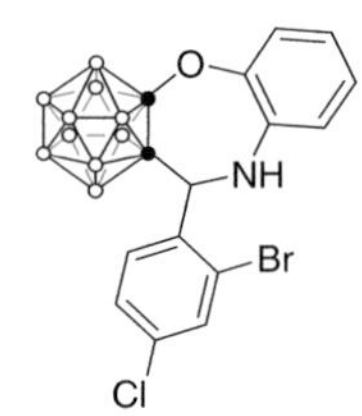

IV-4l, 63%

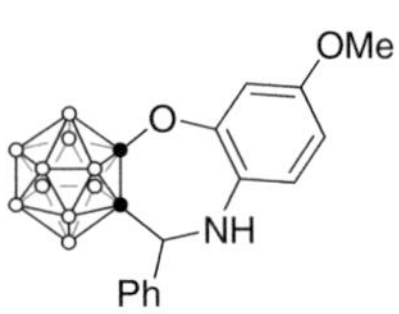

IV-4m, 84%

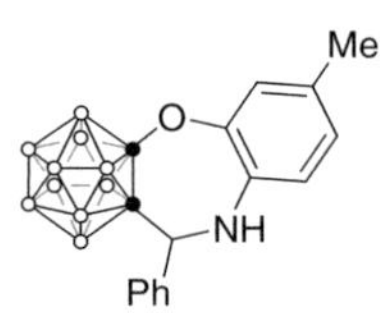

IV-4n, 78%

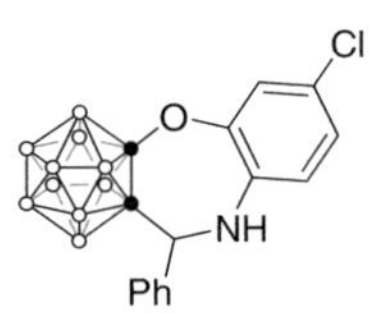

IV-4o, 70%

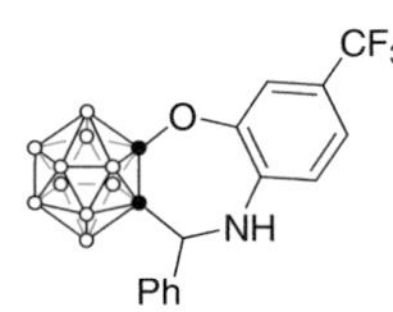

IV-4p, 63%

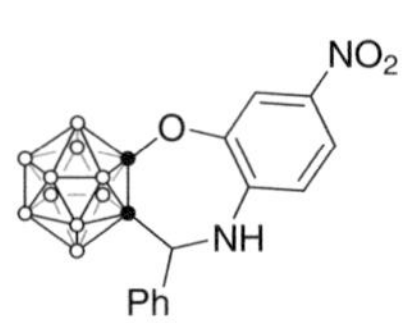

IV-4q, 45%

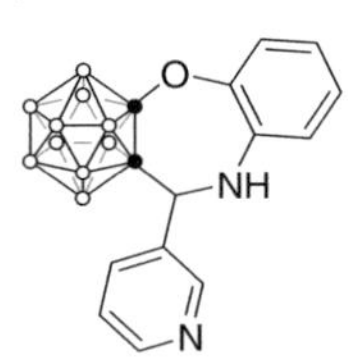

IV-4r, 51%

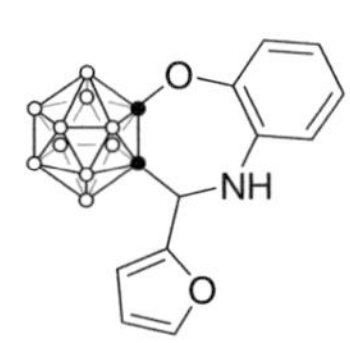

IV-4s, 0%[b]

[a]Reaction conditions: **IV-1** (0.1 mmol), **IV-3** (0.11 mmol), CsF (2.0 mmol), solvent (3 mL). Yields of isolated products

[b]*o*-Carborane was the only isolated product

For nitrone **IV-3r** featuring a 3-pyridyl group, the desired cycloadduct was obtained in 51 % isolated yield. Unfortunately, nitrone **IV-3s**, which contains a 2-furyl group, was completely unreactive. This result may be ascribed to the side reaction of the furan ring with *o*-carboryne [14b, 15].

4.3 Mechanistic Studies

To explore the mechanistic aspects on this unexpected [5 + 2] cycloaddition reaction, several control experiments were conducted (Scheme 4.5). When the *N*-phenyl ring of nitrone is replaced by an alkyl group, for example, **IV-3t**, such [5 + 2] cycloaddition product was not formed. Instead, a formal [3 + 2] cycloaddition took place, similar to the reaction of benzyne with nitrones (Eq. 4.1) [24].

The reaction of **IV-1** with deuterated nitrone **IV-3a-d$_5$** in THF-d_8 was also investigated (Eqs. 4.2–4.3). The desired product **IV-4a-d$_5$** was obtained with more than 99 % deuterium incorporation into the N–H bond based on the analysis of the crude ^{1}H NMR of the reaction mixture (Figs. 4.3, 4.4, and 4.5). On the other hand,

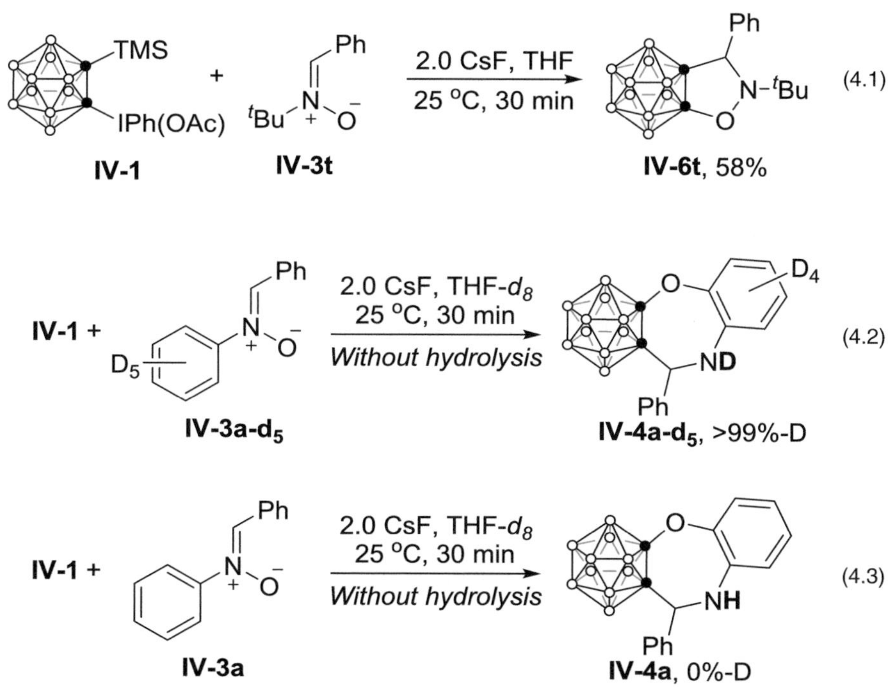

Scheme 4.5 Control experiments

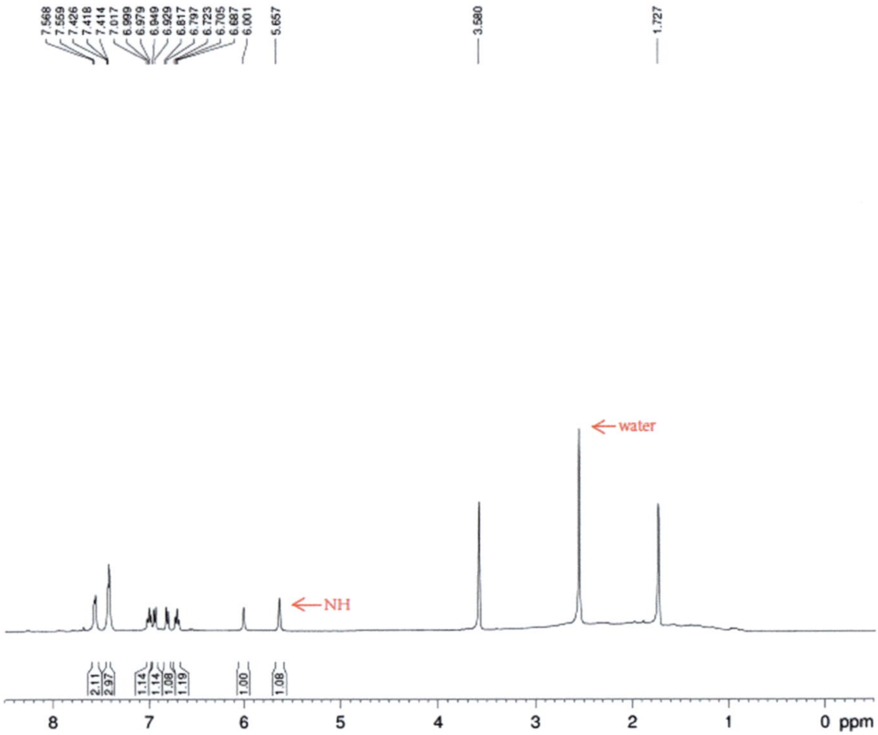

Fig. 4.3 ^{1}H NMR of **IV-4a** in THF-d_8 (for reference)

the reaction of **IV-1** with **IV-3a** in THF-d_8 did not provide any deuterated product, suggesting that the D (H) incorporated into the N–H bond of the product comes from the nitrone substrate (Scheme 4.5).

On the basis of the experimental and theoretical data, a plausible mechanism was proposed in Scheme 4.6. A [3 + 2] cycloaddition between diaryl nitrone **IV-3** and *o*-carboryne **IV-2** generated from precursor **IV-1** in the presence of fluoride affords an intermediate **IV-A** [24e]. Heterolytic cleavage of the N–O bond gives another intermediate **IV-B** in which the negative charge of the oxygen is stabilized by the electron-deficient *o*-carborane cage via *exo*-π bonding [26] and the nitrenium ion is stabilized by the aryl substituent [27, 28]. A following nucleophilic attack of the oxygen anion on the *N*-phenyl ring provides an intermediate **IV-C**, which quickly undergoes rearomatization to furnish the formal [5 + 2] cycloaddition product.

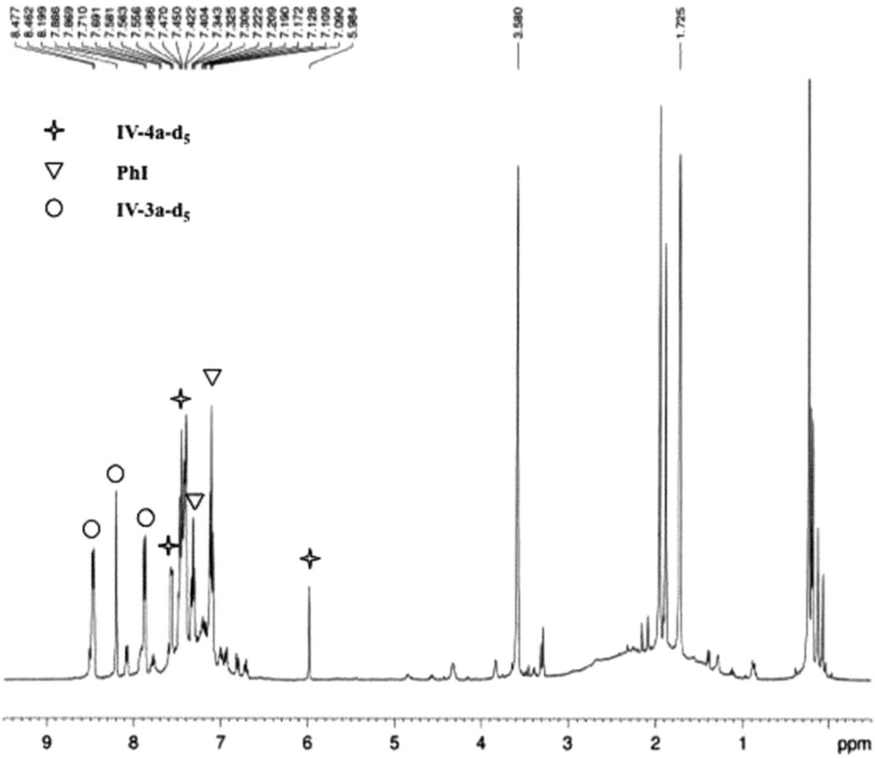

Fig. 4.4 Crude ^{1}H NMR of reaction mixture in THF-d_8 (NMR tube, 60 °C, 60 min)

When the substituent R is more electron-withdrawing, the cleavage of the N–O bond is less favored compared to the electron-donating ones as the aryl group on the nitrogen cannot stabilize the positive charge effectively; thus, lower yield is obtained (Table 4.2, **IV-4o–q**).

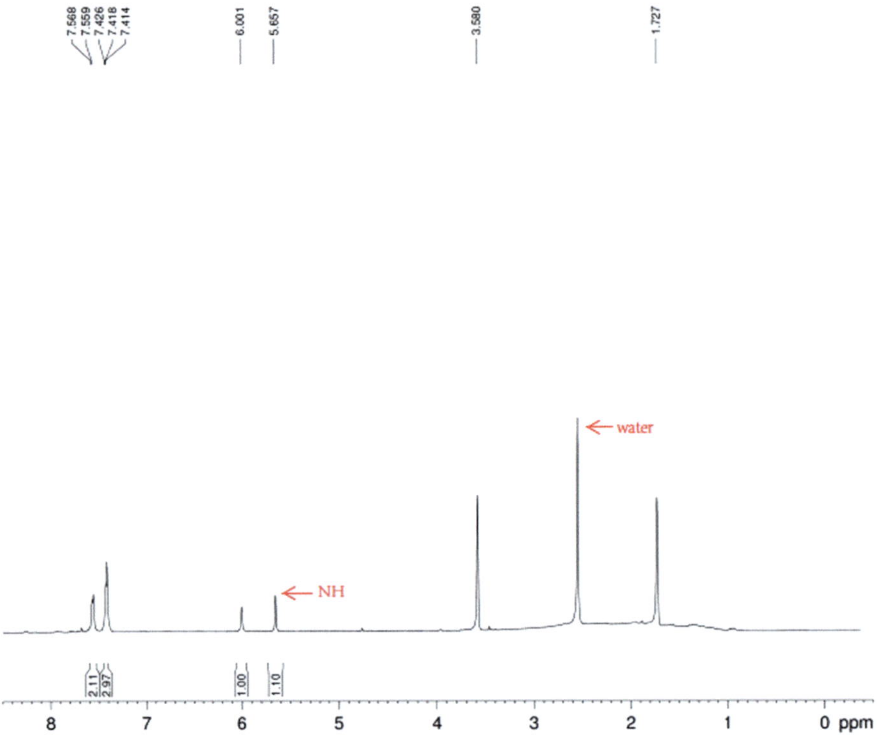

Fig. 4.5 ^{1}H NMR of **IV-4a-d$_4$** (D/H exchange of **IV-4a-d$_5$** with H_2O) in THF-d$_8$

4.4 Summary

In summary, the reaction of diarylnitrones with *o*-carboryne has been described. Precursors for the generation of *o*-carboryne are found to have a dramatic influence on the reaction outcome. In general, nucleophilic addition reaction occurs for *o*-carboryne derived from 1-X-2-lithio-*o*-carborane (X = I, Br). On the other hand, when a neutral precursor, phenyl[*o*-(trimethylsilyl)carboranyl]iodonium acetate, was utilized as *o*-carboryne source, an unexpected [5 + 2] cycloaddition takes place to afford carborane-fused seven-membered heterocycles. In this process, nitrones serve as five-atom building units, which are very rarely discussed in the literature. Mechanistic studies suggest that the N–O bond cleavage in the initially formed [3 + 2] cycloadducts, followed by an intramolecular nucleophilic addition, and rearomatization leads to the final products. This study provides new insight into the chemistry of nitrones as well as *o*-carboryne, which may also find applications in the synthesis of carborane-fused heterocycles.

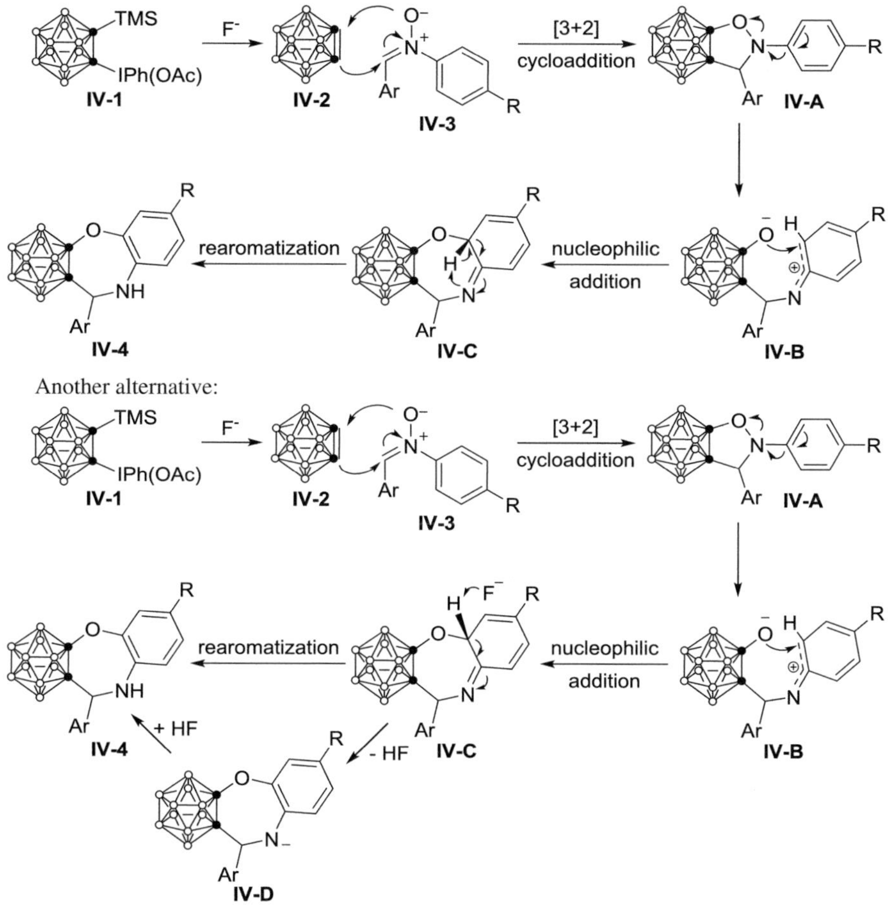

Scheme 4.6 Proposed reaction pathway

References

1. (a) Kobayashi S, Jorgensen KA (2001) Cycloaddition reactions in organic synthesis. Wiley-VCH, Weinheim (b) Padwa A, Pearson WH (2002) Synthetic applications of 1,3-dipolar cycloaddition chemistry toward heterocycles and natural products: the chemistry of heterocyclic compounds. Wiley, New York (c) Fringuelli F, Taticchi A (2002) The Diels–Alder reaction: selected practical methods. Wiley, New York
2. (a) Padwa A (2004) Science of synthesis. Thieme, Stuttgart (b) Padwa A (1984) 1,3-dipolar cycloaddition chemistry. Wiley, New York (c) Frederickson M (1997) Optically active isoxazolidines via asymmetric cycloaddition reactions of nitrones with alkenes: applications in organic synthesis. Tetrahedron 53(2):403–425 (d) Gothelf KV, Hazell RG, Jorgensen KA (1998) Molecular sieve dependent absolute stereoselectivity in asymmetric catalytic 1,3-dipolar cycloaddition reactions. J Org Chem 63(16):5483–5488
3. Qin C, Davies HM (2013) $Rh_2(R\text{-}TPCP)_4$-catalyzed enantioselective [3+2]-cycloaddition between nitrones and vinyldiazoacetates J Am Chem Soc 135(39):14516–14519 (b) Song GY,

Chen D, Su Y et al (2011) Isolation of azomethine ylides and their complexes: Iridium(iii)-mediated cyclization of nitrone substrates containing alkynes. Angew Chem Int Ed 50 (34):7791–7796 (c) Stanley LM, Sibi MP (2008) Enantioselective copper-catalyzed 1,3-dipolar cycloadditions. Chem Rev 108(8):2887–2902

4. (a) Young IS, Kerr MA (2003) A homo [3+2] dipolar cycloaddition: The reaction of nitrones with cyclopropanes. Angew Chem Int Ed 42 (26):3023–3026 (b) Sibi MP, Ma ZH, Jasperse CP (2005) Enantioselective addition of nitrones to activated cyclopropanes. J Am Chem Soc 127(16):5764–5765 (c) Kang YB, Sun XL, Tang Y (2007) Highly enantioselective and diastereoselective cycloaddition of cyclopropanes with nitrones and its application in the kinetic resolution of 2-substituted cyclopropane-1,1-dicarboxylates. Angew Chem Int Ed 46 (21):3918–3921
5. (a) Liu F, Yu Y, Zhang J (2009) Highly substituted furo[3,4-d][1,2]oxazines: gold-catalyzed regiospecific and diastereoselective 1,3-dipolar cycloaddition of 2-(1-alkynyl)-2-alken-1-ones with nitrones. Angew Chem Int Ed 48(30):5505–5508 (b) Liu F, Qian D, Li L et al (2010) Diastereo- and enantioselective gold(i)-catalyzed intermolecular tandem cyclization/[3 +3]cycloadditions of 2-(1-alkynyl)-2-alken-1-ones with nitrones. Angew Chem Int Ed 49 (37):6669–6672
6. (a) Wang X, Xu X, Zavalij PY et al (2011) Asymmeric formal [3+3]-cycloaddition reactions of nitrones with electrophilic vinylcarbene intermediates. J Am Chem Soc 133(41):16402–16405 (b) Qian Y, Xu XF, Wang X et al (2012) Rhodium(ii)- and copper(ii)-catalyzed reactions of enol diazoacetates with nitrones: Metal carbene versus lewis acid directed pathways. Angew Chem Int Ed 51(24):5900–5903 (c) Wang X, Abrahams QM, Zavalij PY et al (2012) Highly regio- and stereoselective dirhodium vinylcarbene induced nitrone cycloaddition with subsequent cascade carbenoid aromatic cycloaddition/n-o cleavage and rearrangement. Angew Chem Int Ed 51(24):5907–5910
7. Yang WB, Wang T, Yu Y et al (2013) Nitrones as trapping reagents of alpha, beta-unsaturated carbene intermediates: [1,2]oxazino[5,4-b]indoles by a platinum-catalyzed intermolecular [3 +3] cycloaddition. Adv Synth Catal 355(8):1523–1528
8. Shintani R, Park S, Duan WL et al (2007) Palladium-catalyzed asymmetric [3+3] cycloaddition of trimethylenemethane derivatives with nitrones. Angew Chem Int Ed 46 (31):5901–5903
9. (a) Shintani R, Murakami M, Hayashi T (2007) Gamma-methylidene-delta-valerolactones as a coupling partner for cycloaddition: palladium-catalyzed [4+3] cycloaddition with nitrones. J Am Chem Soc 129(41):12356–12357 (b) Stevens AC, Palmer C, Pagenkopf BL (2011) The formal [4+3] cycloaddition between donor-acceptor cyclobutanes and nitrones. Org Lett 13 (6):1528–1531
10. (a) Bai Y, Fang J, Ren J et al (2009) Highly diastereoselective gold- or copper-catalyzed formal [4+3] cycloaddition of 1-(1-alkynyl) cyclopropyl ketones and nitrones. Chem-Eur J 15 (36):8975–8978 (b) Zhang Y, Liu F, Zhang J (2010) Catalytic regioselective control in the diastereoselective 1,3-dipolar cycloaddition reactions of 1-(1-alkynyl)cyclopropyl ketones with nitrones. Chem-Eur J 16(21):6146–6150
11. Gawade SA, Bhunia S, Liu RS (2012) Intermolecular gold-catalyzed diastereo-and enantioselective [2+2+3] cycloadditions of 1, 6-enynes with nitrones. Angew Chem Int Ed 124(31):7835–7838
12. Wang C, Wang D, Yan H et al (2014) Rhodium-catalyzed cyclization of diynes with nitrones: a formal [2+2+5] approach to bridged eight-membered heterocycles. Angew Chem Int Ed 53 (44):11940–11943
13. (a) Qiu Z, Wang SR, Xie Z (2010) Nickel-catalyzed regioselective [2+2+2] cycloaddition of carboryne with alkynes. Angew Chem Int Edit 49(27):4649–4652 (b) Qiu Z, Xie Z (2014) Generation and reactivity of o-carborynes. Dalton Trans 43(13):4925–4934
14. (a) Ghosh T, Gingrich HL, Kam CK et al (1991) Reactions of 1,2-dehydro-ortho-carborane with dienes. J Am Chem Soc 113(4):1313–1318 (b) Gingrich HL, Huang Q, Morales AL et al (1992) Reactions of enamines with dehydro aromatic compounds. J Org Chem 57(14): 3803–3806

15. Lee T, Jeon J, Song KH et al (2004) Generation and trapping reaction of an efficient 1,2-dehydrocarborane precursor, phenyl[o-(trimethylsilyl)carboranyl]iodonium acetate. Dalton Trans 6:933–937
16. Atkins JH, Ho DM, Jones M (1996) A new source of 1,2-dehydro-o-carborane. Tetrahedron Lett 37(40):7217–7220
17. Wang SR, Xie Z (2012) Reaction of carboryne with alkylbenzenes. Organometallics 31 (8):3316–3323
18. BarnettThamattoor L, Zheng GX, Ho DM et al (1996) Reactions of 1,2-dehydro-o-carborane with thiophenes. Cycloadditions and an easy synthesis of "benzo-o-carboranes". Inorg Chem 35(25):7311–7315
19. Wang SR, Xie Z (2012) Reaction of carboryne with styrene and its derivatives. Tetrahedron 68 (26):5269–5278
20. Kuehn ME (1962) The arylation of enamines. J Am Chem Soc 84(5):837–847
21. (a) Cunningham RJ, Bian NY, Jones M (1994) Reaction of 1,2-dehydro-o-carborane with acetylenes—ene reactions and cycloaddition to give an analog of benzocyclobutadiene. Inorg Chem 33(22):4811–4812 (b) Ho DM, Cunningham RJ, Brewer JA et al (1995) Reaction of 1,2-dehydro-o-carborane with acetylenes—synthesis and structure of a carborane analog of benzocyclobutadiene. Inorg Chem 34(21):5274–5278
22. Wang SR, Qiu Z, Xie Z (2010) Insertion of carboryne into aromatic rings: formation of cyclooctatetraenocarboranes. J Am Chem Soc 132(29):9988–9989
23. Xu W (2014) M.Phil. Thesis. The Chinese University of Hong Kong, Hong Kong
24. (a) Dai M, Wang Z, Danishefsky SJ (2008) A novel alpha,beta-unsaturated nitrone-aryne [3 +2] cycloaddition and its application in the synthesis of the cortistatin core. Tetrahedron Lett 49(47):6613–6616 (b) Wu K, Chen Y, Lin Y et al (2010) Cycloaddition of nitrones with arynes generated from benzobisoxadisilole or 2,3-naphthoxadisilole. Tetrahedron 66(3): 578–582 (c) Lu C, Dubrovskiy AV, Larock RC (2012) Synthesis of benzisoxazolines by the coupling of arynes with nitrones. J Org Chem 77(5):2279–2284
25. (a) Tomita H, Luu H, Onak T (1991) Cage opening of parent closo cage carboranes with fluoride-ion—formation of [5-f-nido-2,4-c2b4h6]-. Inorg Chem 30(4):812–815 (b) Fox MA, Gill WR, Herbertson PL et al (1996) Deboronation of C-substituted ortho- and meta-closo-carboranes using "wet" fluoride ion solutions. Polyhedron 15(4):565–571 (c) Getman TD (1998) Investigation of potassium fluoride supported on alumina in the deboronation of o-carborane. Inorg Chem 37(13):3422–3423 (d) Yoo J, Hwang J. W, Do, Y (2001) Facile and mild deboronation of o-carboranes using cesium fluoride. Inorg Chem 40 (3):568–57
26. (a) Boyd LA, Clegg W, Copley RCB et al (2004) Exo-pi-bonding to an ortho-carborane hypercarbon atom: systematic icosahedral cage distortions reflected in the structures of the fluoro-, hydroxy- and amino-carboranes, 1-X-2-Ph-1,2$C_2B_{10}H_{10}$ (X=F, OH or NH_2) and related anions. Dalton Trans (17):2786–2799 (b) Getman TD, Knobler CB, Hawthorne MF (1992) Reduction of biscarborane—molecular-structures of $[(15\text{-crown-}5)_3Na_2](C_2B_{10}H_{11})_2$ and $[P(C_6H_5)_3CH_3]_2$[mu-9,10-CH-(mu-9′,10′-CH-nido-7′-$CB_{10}H_{11}$)-nido-7-$CB_{10}H_{11}$]. Inorg Chem 31(1):101–105 (c) Oliva JM, Allan NL, Schleyer PV et al (2005) Strikingly long C–C distances in 1,2-disubstituted ortho-carboranes and their dianions. J Am Chem Soc 127(39): 13538–13547
27. (a) Gassman PG (1970) Nitrenium ions. Acc Chem Res 3(1):26–33 (b) Falvey DE, Gudmundsdottir AD (2013) Nitrenes and Nitrenium Ions. Wiley, New Jersey
28. (a) Cramer CJ, Dulles FJ, Falvey DE (1994) Ab-initio characterization of phenylnitrenium and phenylcarbene—remarkably different properties for isoelectronic species. J Am Chem Soc 116 (21):9787–9788 (b) Winter AH, Falvey DE, Cramer CJ et al (2007) Benzylic cations with triplet ground states: computational studies of aryl carbenium ions, silylenium ions, nitrenium ions, and oxenium ions substituted with meta pi donors. J Am Chem Soc 129(33): 10113–10119 (c) Sullivan MB, Brown K, Cramer CJ et al (1998) Quantum chemical analysis of para-substitution effects on the electronic structure of phenylnitrenium ions in the gas phase and aqueous solution. J Am Chem Soc 120(45):11778–11783

Chapter 5
1,3-Dehydro-*o*-Carborane: Generation and Reaction with Arenes

Abstract This chapter presents the generation of 1,3-dehydro-*o*-carborane, a previously unknown intermediate featuring a cage C–B multiple bond. This reactive intermediate is best described as a resonance hybrid of a bonding form and a zwitterionic form, and it can undergo [4 + 2] reaction as well as aromatic ene reaction with arenes. This serves as a new methodology for simultaneous functionalization of both cage carbon and boron vertices.

5.1 Introduction

Arynes, which are formally derived from aromatic rings by abstraction of two hydrogen atoms, have been a focus of organic chemistry for 100 years, and nowadays, arynes, heteroarynes, and related intermediates are among the most intensively studied intermediates in the literature (Fig. 5.1) [1].

During our study on *o*-carboryne chemistry [2], we found that *o*-carboryne can be viewed as a three-dimensional relative of benzyne and they share some common features; on the other hand, *o*-carboryne has its own unique properties due mainly to steric/electronic reasons [2d]. To our surprise, the known arynes and related intermediates including *o*-carboryne all contain a carbon–carbon multiple bond and no example of such intermediate that features a carbon–heteroatom multiple bond has been reported. On the basis of our experience in carborane chemistry, we wonder whether 1,3-dehydro-*o*-carborane [3], which contains a cage C–B multiple bond, could be generated (Fig. 5.1). Indeed, it represents the first example of such intermediate that contains a carbon–heteroatom multiple bond. Moreover, it would also serve as a useful synthon for simultaneous functionalization of both cage carbon and boron vertices. In this section, the generation and reactivity study of this new intermediate will be discussed.

D. Zhao, *Functionalization of Carborane via Carboryne Intermediates*,
Springer Theses, DOI 10.1007/978-981-10-1569-4_5

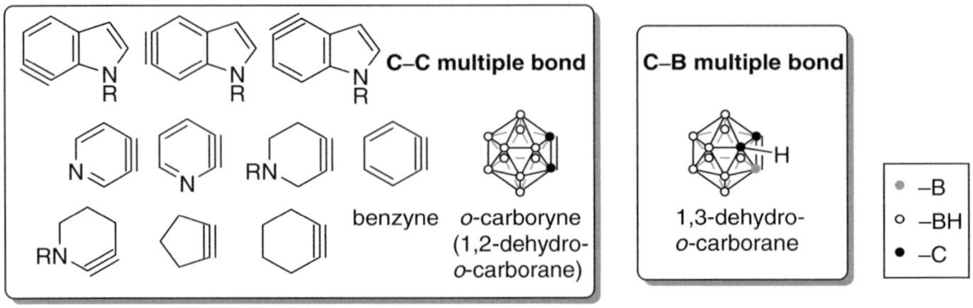

Fig. 5.1 Reported examples of arynes and related intermediates

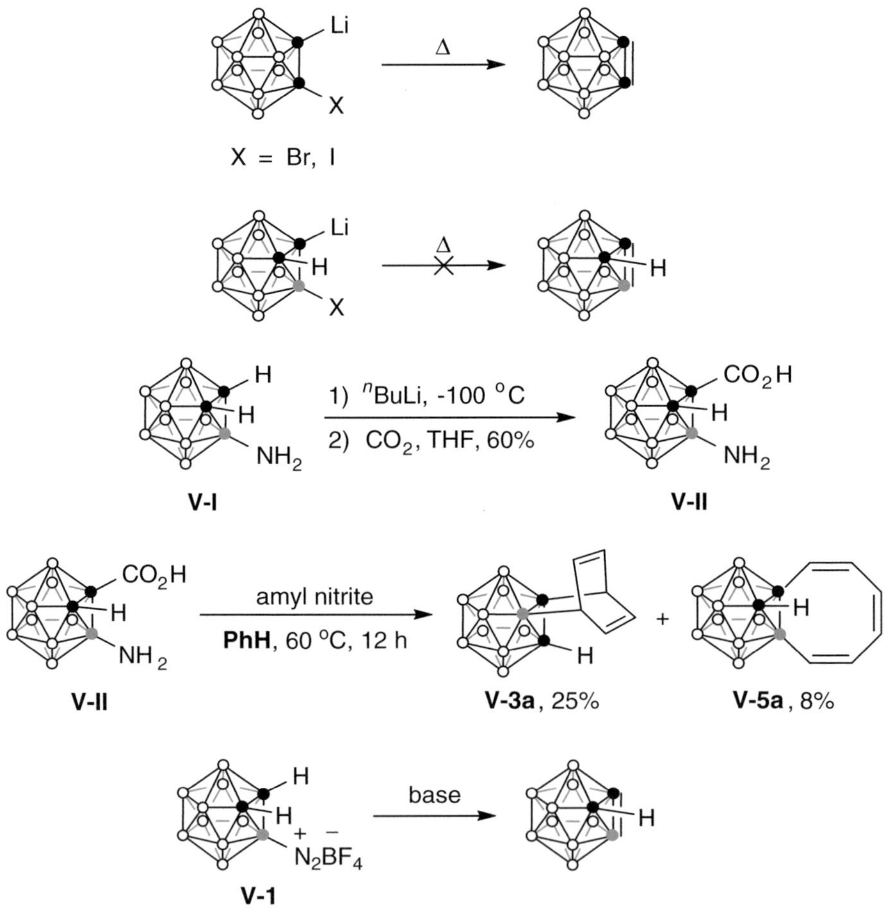

Scheme 5.1 Attempt to generate 1,3-dehydro-*o*-carborane

5.2 Generation of 1,3-Dehydro-*o*-Carborane: Precursor Design

In order to generate 1,3-dehydro-*o*-carborane, we speculated that it might be generated in a similar manner to that of *o*-carboryne, which can be produced from 1-X-2-Li-*o*-$C_2B_{10}H_{10}$, via LiX salt elimination (X = Br, I) [2a, c]. However, attempts to obtain 1,3-dehydro-*o*-carborane from 1-Li-3-X-*o*-$C_2B_{10}H_{10}$ (X = Br, I) by LiX elimination failed since 1-Li-3-X-*o*-$C_2B_{10}H_{10}$ was thermally stable, even under forced reaction conditions, owing to a very strong B–X bond (Scheme 5.1) [4].

Inspired by earlier benzyne generation from anthranilic acids [5], we turned our attention to the synthesis of 1-hydroxycarbonyl-3-amino-*o*-carborane (**V-II**). As shown in Scheme 5.1, **V-II** was prepared in 60 % isolated yield by the treatment of 3-amino-*o*-carborane (**V-I**) with 4.5 equiv of *n*-butyllithium followed by carboxylation with CO_2 (g). Diazotization of **V-II** with amyl nitrite in benzene (**V-2a**) at 60 °C afforded the [4 + 2] cycloadduct **V-3a** and 1,3-cyclooctatetraenocarborane **V-5a** in 25 and 8 % yield, respectively. The molecular structures of **V-3a** and **V-5a** were confirmed by various spectroscopic techniques as well as single-crystal X-ray analyses (Figs. 5.2 and 5.3). The unexpected product **V-5a** is probably derived from a thermal [3,3] sigmatropic rearrangement of the [2 + 2] cycloaddition intermediates based on our previous study [6].

To our disappointment, diazotization of **V-II** in other solvents (THF, ether, and CH_3CN) was surprisingly difficult. On the other hand, comparing to the reactivity of *o*-carboryne toward benzene (24 % yield) [7], the efficiency was not satisfactorily enough.

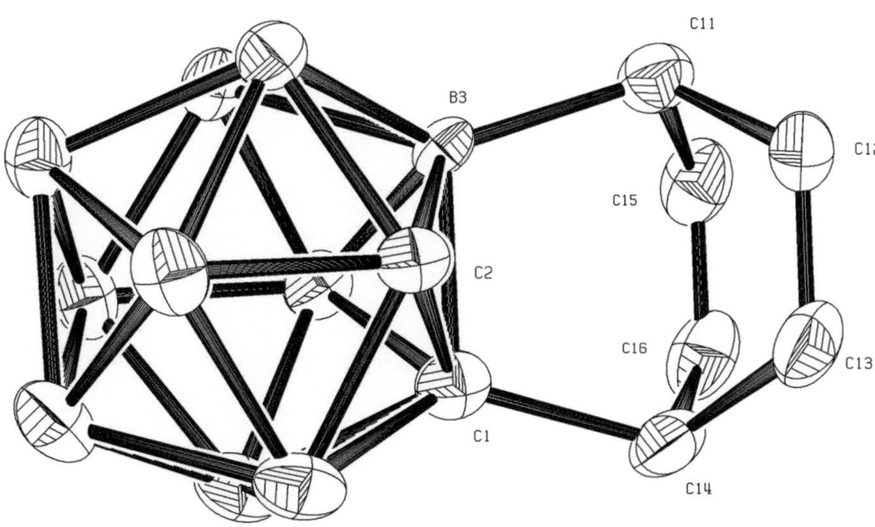

Fig. 5.2 Molecular structure of **V-3a**

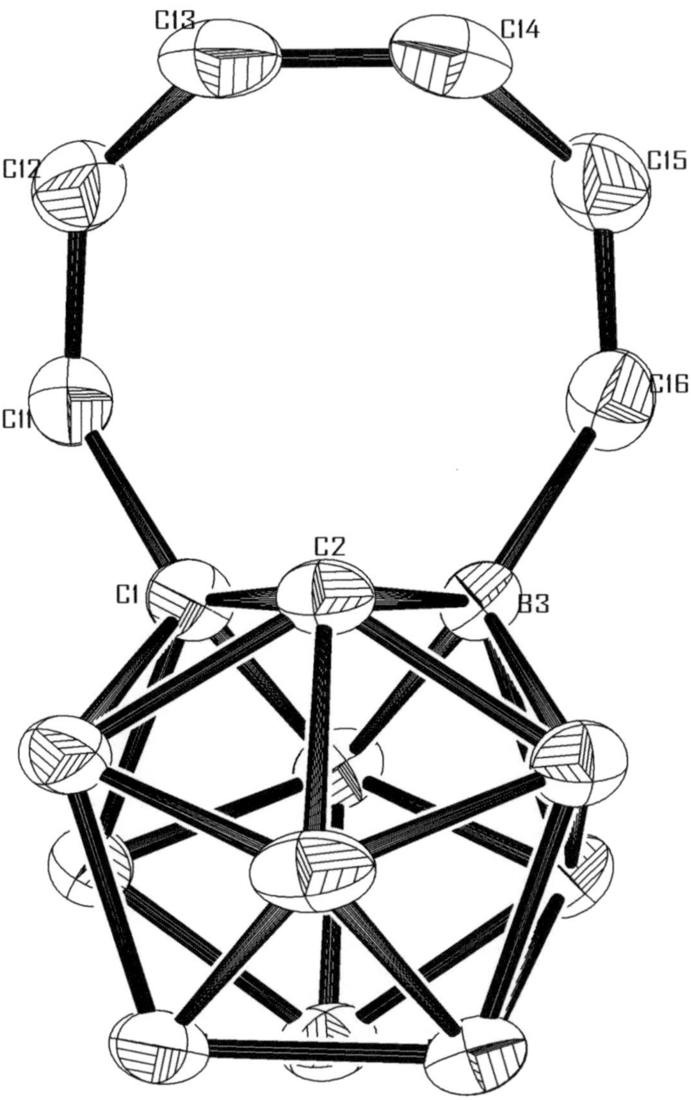

Fig. 5.3 Molecular structure of **V-5a**

5.3 3-($N_2^+BF_4^-$)-*o*-$C_2B_{10}H_{11}$: A More Efficient Precursor

In view of the properties of diazonium salt of carboranes [8], we thought that 3-($N_2^+BF_4^-$)-*o*-$C_2B_{10}H_{11}$ (**V-1**) may serve as a good precursor of 1,3-dehydro-*o*-carborane as dinitrogen is an excellent leaving group after deprotonation of cage C–H (Scheme 5.1). Precursor **V-1** was prepared in 70 % isolated yield by the treatment of 3-amino-*o*-carborane **V-I** with 1.2 equiv of in situ-generated nitrosyl

Table 5.1 Screening of reaction conditions

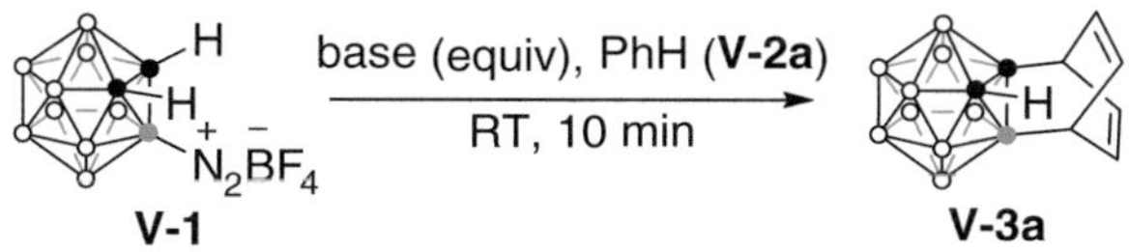

Entry	Base (equiv)	**3a** (%)[a]
1	^{n}BuLi (1.0)	38
2	LDA (1.0)	72
3	NaH (1.0)	55
4	$NaNH_2$ (1.0)	61
5	LiHMDS (1.0)	66
6	LDA (2.0)	72
7[b]	LDA (1.0)	72

[a]Isolated yields
[b]The reaction was performed in the dark. LiHMDS = lithium bis(trimethylsilyl)amide

fluoride in the presence of boron trifluoride [9]. It was noted that the stability of **V-1** is dependent upon the counterion used and BF_4^- offers the highest thermal stability of the salt among the anions examined such as PF_6^- and Cl^-.

To test our hypothesis, a benzene suspension of **V-1** was treated with 1.0 equiv of ^{n}BuLi at room temperature for 10 min to give the expected [4 + 2] cycloaddition product **V-3a** in 38 % isolated yield (Table 5.1, entry 1) [10]. The low yield was due to the formation of 3-^{n}Bu-o-$C_2B_{10}H_{10}$ generated by the nucleophilic attack of ^{n}BuLi [11]. When non-nucleophilic base lithium diisopropylamide (LDA) was used, **V-3a** was obtained in 72 % isolated yield (Table 5.1, entry 2). Other less nucleophilic bases gave relatively lower yields (Table 5.1, entries 3–5). Increasing the amount of base did not improve the yield (Table 5.1, entry 6). The reaction also proceeded very well in the dark (Table 5.1, entry 7). These results suggest that **V-1** is a very good precursor of 1,3-dehydro-o-carborane and its reaction with benzene is much more efficient than that of o-carboryne [6].

5.4 Reactivity of 1,3-Dehydro-*o*-Carborane Toward Arenes

Under the optimal reaction condition (Table 5.1, entry 2), various arenes with different substituents such as trimethylsilyl, halo, and alkyls were examined. The results are given in Table 5.2. In general, the reaction efficiency of 1,3-dehydro-o-carborane with arenes was much higher than that of o-carboryne. For methyl-substituted arenes **V-2i–m**, the corresponding aromatic ene reaction product

Table 5.2 Reaction of arenes with 1,3-dehydro-*o*-carborane[a]

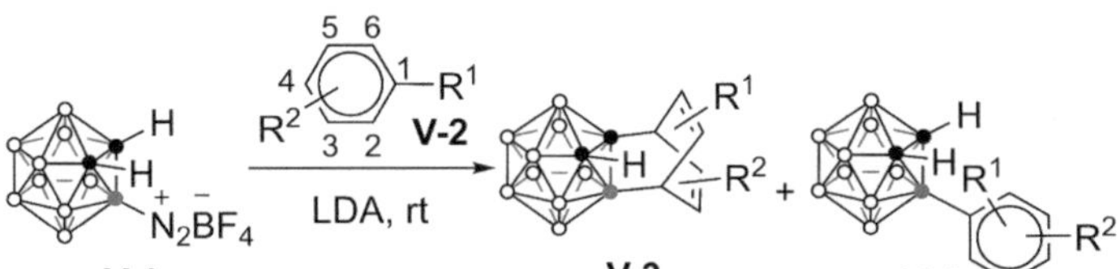

Entry	R^1/R^2 (**V-2**)	Yield (%)[b]	
		V-3	**V-4**
1	H/H (**V-2a**)	72	–
2	F/2,3,4,5,6-F_5 (**V-2b**)	25[c]	–
3	F/H (**V-2c**)	64	–
4	Cl/H (**V-2d**)	67	–
5	Cl/2-Cl (**V-2e**)	55	–
6	CF_3/H (**V-2f**)	53	–
7	TMS/H (**V-2g**)	52	–
8	^{t}Bu/H (**V-2h**)	67	–
9	Me/H (**V-2i**)	55	27
10	Me/2-Me (**V-2j**)	47	39
11	Me/3-Me (**V-2k**)	26	53
12	Me/4-Me (**V-2l**)	39	41
13	Me/3,5-Me_2 (**V-2m**)	24	69

[a]Conditions: LDA (1.0 equiv), arenes **V−2** (40.0 equiv), room temperature, 10 min
[b]Isolated yields
[c]The low isolated yield resulted from the high volatility of **V-3b**

[12] 3-aryl-*o*-carborane (**V-4**) was also isolated and became a major one in cases of **V-2k** and **V-2m** (Table 5.2, entries 11 and 13). The molecular structures of [4 + 2] cycloadducts **V-3h, V-3l, V-3m**, and aromatic ene reaction products **V-4j** and **V-4m** are shown in Figs. 5.4, 5.5, 5.6, 5.7 and 5.8.

Since carbon and boron atoms are different in nature and the cage C–B bonds are polarized, the reactive intermediate 1,3-dehydro-*o*-carborane is best described as a resonance hybrid of both bonding and zwitterionic forms (Scheme 5.2a). Thus, the regioselectivity of the [4 + 2] cycloaddition of 1,3-dehydro-*o*-carborane with substituted arenes is governed by electronic factors although steric factors may also play a role, which is similar to that of hetero-Diels–Alder reactions [13] or the reactions of distorted arynes [14]. Scheme 5.2b shows both electronically favored and disfavored fashions of [4 + 2] cycloaddition of 1,3-dehydro-*o*-carborane with substituted arenes.

For arenes having more evenly distributed electron density, two or more [4 + 2] cycloaddition fashions were observed [10], whereas in the case of **V-2f**, only one [4 + 2] cycloaddition fashion was observed, probably owing to the strong inductive effect imparted by the CF_3 group. As 1,3-dehydro-*o*-carborane is an unsymmetrical

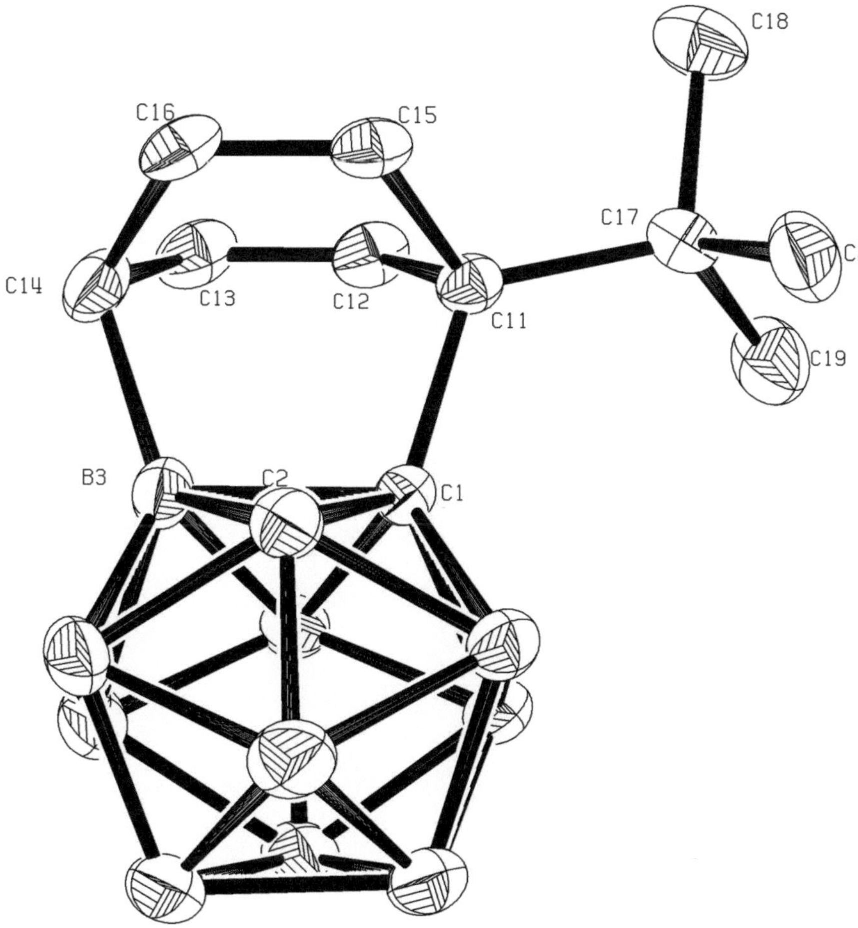

Fig. 5.4 Molecular structure of **V-3h**

species, it is expected that its reaction with substituted arenes affords structural isomers of the products. For instance, two structural isomers were isolated in the reaction of **V-2f** mentioned above (Scheme 5.2c).

Remarkably different from that of *o*-carboryne, the reactions with toluene and its derivatives **V-2i,j,k,l,m** also gave the aromatic ene reaction product [12] **V-4** in 27–69 % isolated yields in addition to the [4 + 2] cycloadducts. The ratio of ene product to [4 + 2] cycloadducts increased with the number of the benzylic C–H bonds (entries 9–13, Table 2). In case of the reaction of **V-2m**, an intermediate of such aromatic ene reaction was observed from the ^{1}H NMR spectrum of the crude product mixture and it rearomatized to **V-4m** within 30 min. It is noteworthy that such aromatic ene reaction proceeded with excellent regioselectivity, giving only cage B(3)-substituted products 3-aryl-*o*-$C_2B_{10}H_{11}$ (**V-4**). This can be ascribed to the zwitterionic nature of 1,3-dehydro-*o*-carborane as illustrated in Scheme 5.3.

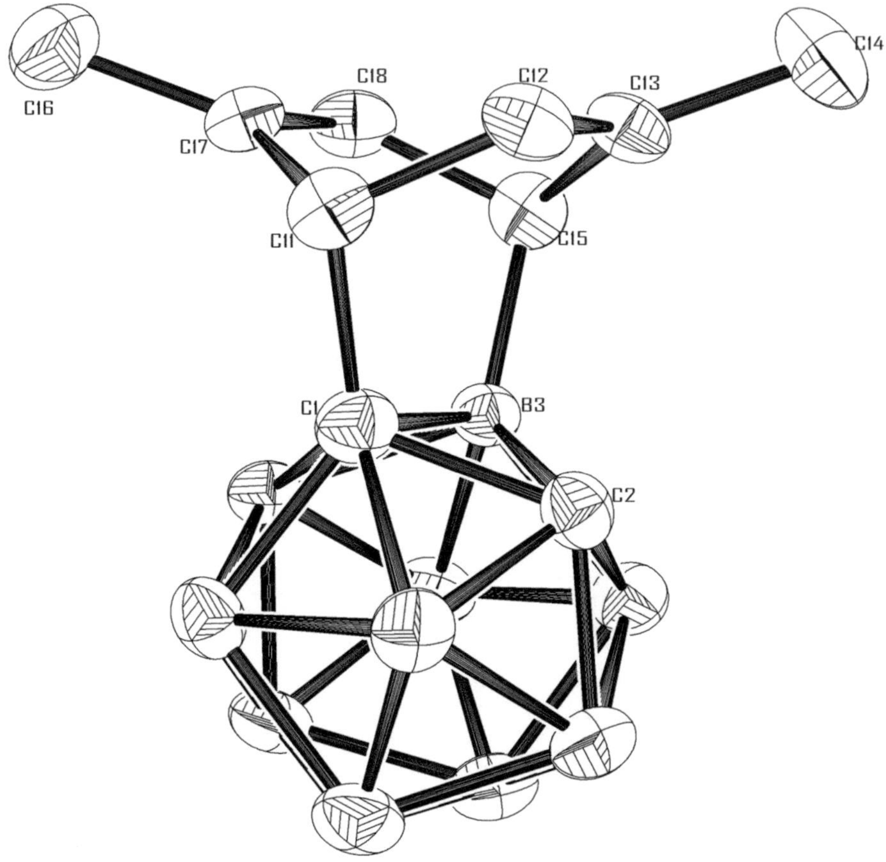

Fig. 5.5 Molecular structure of **V-3l**

5.5 Thermal Rearrangement

It is documented that the [4 + 2] cycloadducts of *o*-carboryne with arenes can undergo thermal retro-Diels–Alder reaction to produce the homo-Diels–Alder products in almost quantitative yield [7]. To explore the possibility of cycloadducts as a new source of 1,3-dehydro-*o*-carborane via retro-Diels–Alder reaction, pyrolysis of the Diels–Alder adducts of 1,3-dehydro-*o*-carborane was conducted. Unprecedentedly, heating of a solid of the [4 + 2] cycloadduct **V-3a** at 250 °C for 6 h in a sealed tube gave a 2:1 mixture of **V-4a** and **V-5a** in 80 % yield (Scheme 5.4). Compound **V-5a** was completely converted to **V-4a** at 250 °C for 6 h, indicating that **V-4a** is a thermodynamically more stable product. On the other

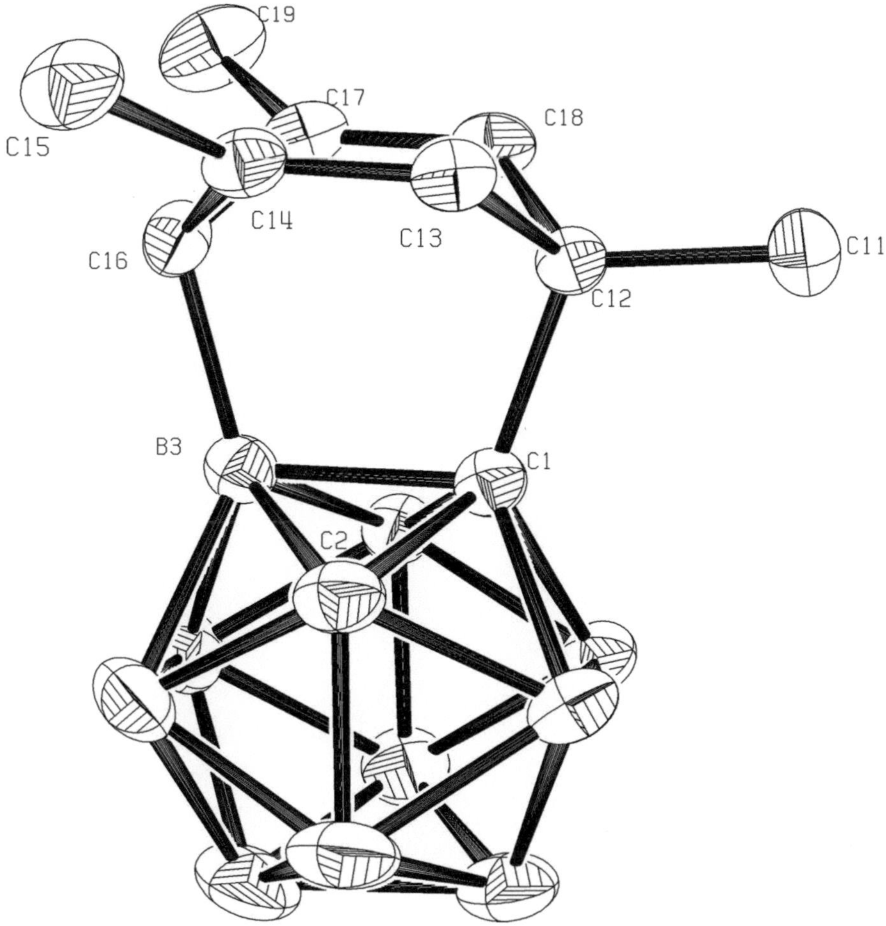

Fig. 5.6 Molecular structure of **V-3m**

hand, heating of mesitylene (**V-2m**) solution of **V-3a** at 250 °C for 6 h in a sealed tube afforded **V-4a** in quantitative yield (Scheme 5.4). Neither the [4 + 2] cycloadduct **V-3m** nor aromatic ene reaction product **V-4m** was observed. These results indicate that the cage B–C bond does not break in the above reactions and they are intramolecular transformations.

To understand these processes, DFT calculations were performed (Fig. 5.9, cage CHs were omitted for clearance). The DFT calculations were carried out with the Gaussian 03 program. Geometry optimizations were performed at the level of B3LYP/6-31+g**. Frequency calculations were made to determine the

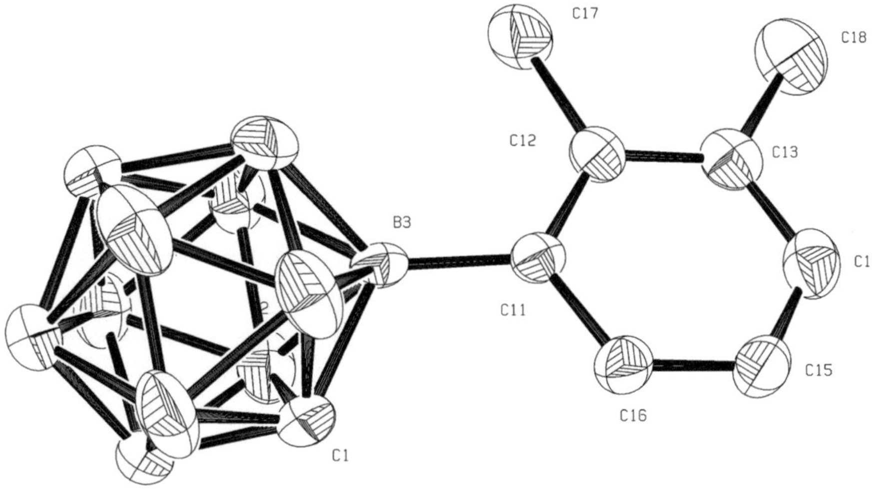

Fig. 5.7 Molecular structure of **V-4j**

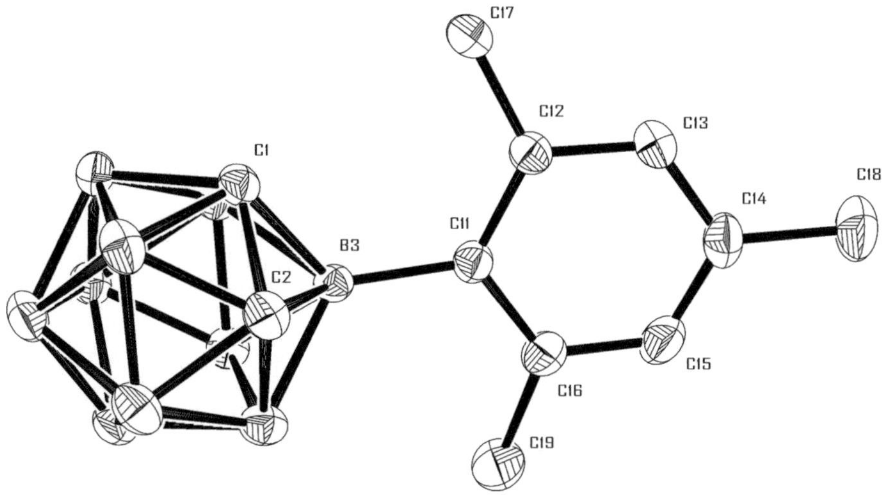

Fig. 5.8 Molecular structure of **V-4m**

characteristics of all stationary points as energy minima or transition states and obtain thermal corrections. Intrinsic reaction coordinates (IRCs) were calculated to confirm that transition states lead to relevant intermediates.

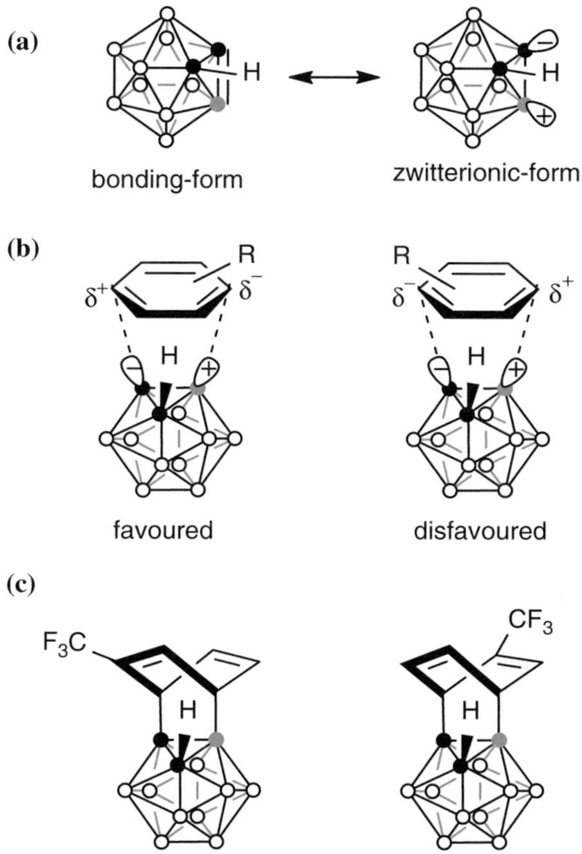

Scheme 5.2 **a** Two resonance forms of 1,3-dehydro-*o*-carborane, **b** [4 + 2] cycloaddition fashions, and **c** two structural isomers generated for **V-2f**

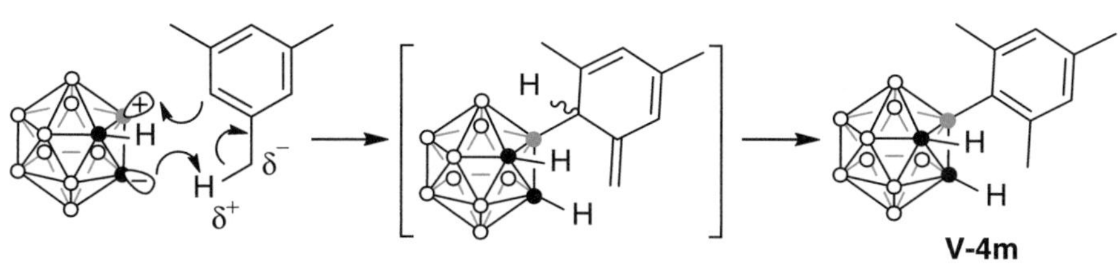

Scheme 5.3 Aromatic ene reaction of 1,3-dehydro-*o*-carborane

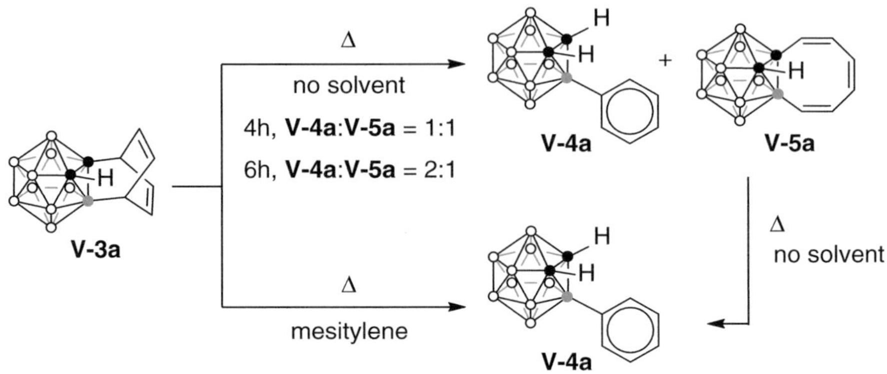

Scheme 5.4 Pyrolysis of cycloadduct **V-3a**

Starting from the [4 + 2] cycloadduct **V-3a**, a heterolytic cleavage of Ccage–C_4 bond via the transition state **V-TS1** (25.7 kcal/mol) generates a zwitterionic intermediate **V-Int2**. Then, a facile rotation of the remaining Bcage-C_1 bond gives the intermediate **V-Int4**, which allows H in C_1H to be orientated toward the bare cage carbon. From **V-Int4**, a proton shift via **Path A** yields product **V-4a** with significant exergonicity. Alternatively, nucleophilic attack of the bare cage carbon on the nearby carbon C_2 via **Path B** gives a formally [2 + 2] cycloadduct **V-Int7**, which further undergoes sigmatropic rearrangement to yield the product **V-5a**. The calculated results reveal that **Path B** leading to the formation of product **V-5a** is kinetically more favorable than **Path A** leading to the formation of product **V-4a** (**V-TS6**: 27.4 kcal/mol vs. **V-TS5**: 31.4 kcal/mol), whereas product **V-4a** is thermodynamically more stable than product **V-5a** by 42.1 kcal/mol. Therefore, with prolonged reaction, the kinetic product **V-5a** can gradually transform to the thermodynamic product **V-4a**. Relevant to the heterolytic cleavage of Ccage–C_4 bond shown in Fig. 5.9, a similar process of Bcage-C_1 bond cleavage cannot be located, probably due to the high energy of C–B bond cleavage.

Accordingly, a reaction mechanism is proposed in Scheme 5.5. Heterolytic cleavage of the cage C–C(sp^3) bond in **V-3a** generates an intermediate **V-A**. Migration of proton in **V-A** gives the thermodynamic product **V-4a**. Alternatively, the formation of cage C–C(sp^2) bond affords another intermediate **V-B** (formally [2 + 2] cycloadduct) that undergoes sigmatropic rearrangement to yield the kinetic product **V-5a**.

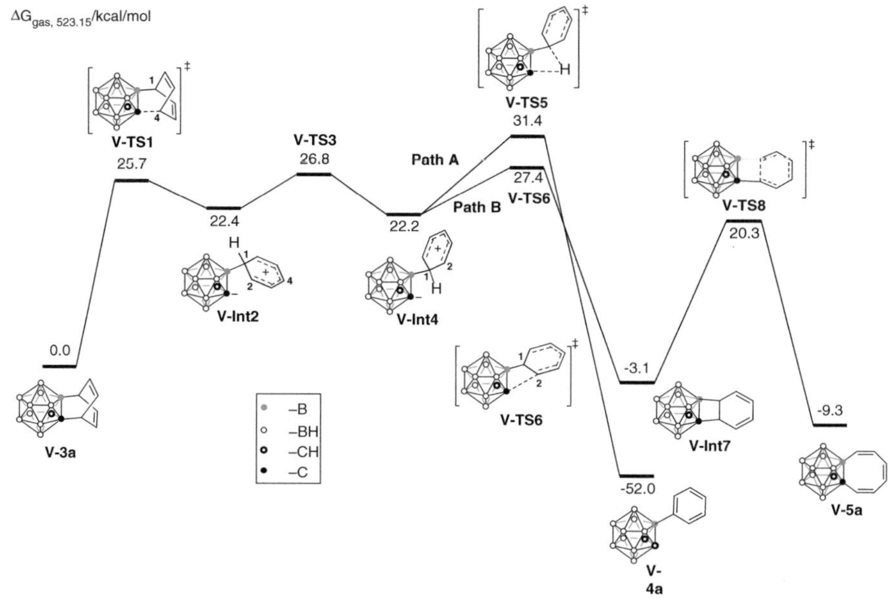

Fig. 5.9 Energy profile ($\Delta G_{gas,523.15}$, kcal/mol) calculated for the pyrolysis of cycloadduct **V-3a**, at the level of B3LYP/6-31+g**

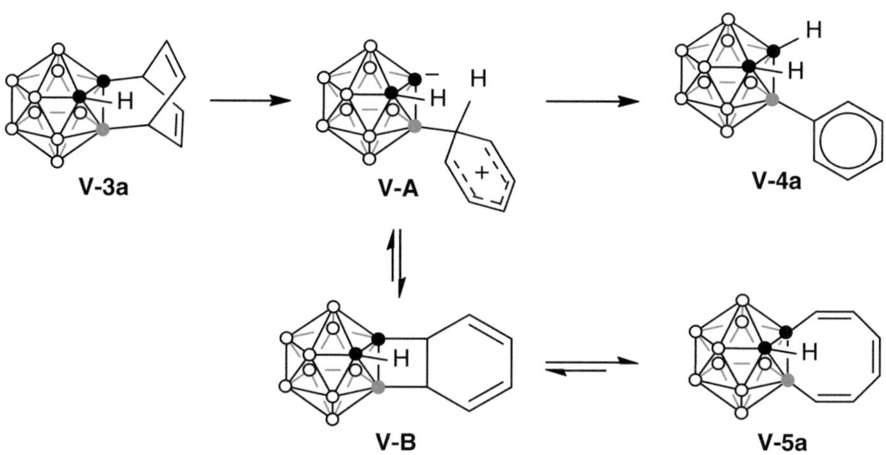

Scheme 5.5 Proposed mechanism for thermal rearrangement

5.6 Summary

We have designed and synthesized an efficient precursor of 1,3-dehydro-*o*-carborane featuring a cage C–B multiple bond. This intermediate represents the first reactive intermediate that contains a carbon–heteroatom multiple bond.

1,3-Dehydro-*o*-carborane undergoes Diels–Alder reaction with various arenes to give [4 + 2] cycloadducts, in which electronic factor governs the regioselectivity. Meanwhile, for arenes bearing benzylic C–H bonds, highly regioselective aromatic ene reaction was also observed with the product ratio of [4 + 2] cycloaddition over ene reaction depending on the number of benzylic protons. These results suggest that 1,3-dehydro-*o*-carborane is best described as a resonance hybrid of a bonding form and a zwitterionic form as shown in Scheme 5.2. Thus, it shares some chemical properties with those of *o*-carboryne; on the other hand, it also has its own unique properties due mainly to the polarized cage C–B bond. The present work demonstrates that 1,3-dehydro-*o*-carborane is a useful synthon for simultaneous cage C,B-functionalization of *o*-carboranes [15].

References

1. (a) Tadross PM, Stoltz BM (2012) A comprehensive history of arynes in natural product total synthesis. Chem Rev 112 (6):3550–3577 (b) Pena D, Perez D, Guitian E (2006) Insertion of arynes into sigma bonds. Angew Chem Int Ed 45 (22):3579–3581 (c) Wenk HH, Winkler M, Sander W (2003) One century of aryne chemistry. Angew Chem Int Ed 42 (5):502–528
2. (a) Qiu Z, Wang SR, Xie Z (2010) Nickel-catalyzed regioselective [2 + 2 + 2] cycloaddition of carboryne with alkynes. Angew Chem Int Edit 49 (27):4649–4652 (b) Qiu Z, Xie Z (2014) Generation and reactivity of o-carborynes Dalton Trans 43(13):4925–4934
3. Qiu Z, Xie Z (2010) Palladium/nickel-cocatalyzed cycloaddition of 1,3-dehydro-*o*-carborane with alkynes. Facile synthesis of c, b-substituted carboranes. J Am Chem Soc 132(45): 16085–16093
4. (a) Cottrell TL (1958) The Strengths of Chemical Bonds, 2nd edn. Butterworth, London (b) Darwent BdeB National Standard Reference Data Series, National Bureau of Standards, Washington (c) Benson SW (1965) III-Bond energies. J Chem Educ 42 (9):502–518 (d) Kerr JA (1966) Bond dissociation energies by kinetic methods Chem Rev 66(5):465–500
5. (a) Friedman L, Logullo FM (1963) Benzynes via aprotic diazotization of anthranilic acids: convenient synthesis of triptycene and derivatives. J Am Chem Soc 85 (10):1549–1549 (b) Friedman L, Logullo FM (1969) Arynes via aprotic diazotization of anthranilic acids. J Org Chem 34 (10):3089–3092
6. Wang SR, Qiu Z, Xie Z (2010) Insertion of carboryne into aromatic rings: formation of cyclooctatetraenocarboranes. J Am Chem Soc 132(29):9988–9989
7. Wang SR, Xie Z (2012) Reaction of carboryne with alkylbenzenes. Organometallics 31 (8):3316–3323
8. Ringstrand B, Kaszynski P (2013) Functionalization of the [closo-1-$CB_9H_{10}]^-$ anion for the construction of new classes of liquid crystals. Acc Chem Res 46(2):214–225
9. Doyle MP, Bryker WJ (1979) Alkyl nitrite-metal halide deamination reactions.6. Direct synthesis of arenediazonium tetrafluoroborate salts from aromatic-amines, tert-butyl nitrite, and boron-trifluoride etherate in anhydrous media. J Org Chem 44(9):1572–1574

10. Zhao D, Zhang J, Xie Z (2014) 1,3-dehydro-*o*-carborane: generation and reaction with arenes. Angew Chem Int Ed 53(32):8488–8491
11. Olid D, Vinas C, Teixidor F (2012) Li^{+}-mediated B-C cross-coupling. Chem-Eur J 18 (41):12936–12940
12. (a) Niu D, Hoye TR (2014) The aromatic ene reaction. Nat Chem 6 (1):34–40 (b) Brinkley JM, Friedman L (1972) Novel ene and insertion reactions of benzyne with alkylbenzenes. Tetrahedron Lett (40):4141–4142 (c) Tabushi I, Yamada H, Yoshida Z et al (1977) Reactions of benzyne with substituted benzenes. B Chem Soc Jpn 50 (1):285–290
13. (a) Boger DL, Weinreb SM (1987) Hetero Diels–Alder methodology in organic synthesis. Academic Press, San Diego (b) Tietze LF, Kettschau G (1997) Hetero diels-alder reactions in organic chemistry. Top Curr Chem 189:1–120
14. (a) Goetz AE, Garg NK (2013) Regioselective reactions of 3,4-pyridynes enabled by the aryne distortion model. Nat Chem 5 (1):54–60 (b) Bronner SM, Mackey JL, Houk KN et al (2012) Steric effects compete with aryne distortion to control regioselectivities of nucleophilic additions to 3-silylarynes. J Am Chem Soc 134 (34):13966–13969 (c) Cheong PHY, Paton RS, Bronner SM et al (2010) Indolyne and aryne distortions and nucleophilic regioselectivites. J Am Chem Soc 132 (4):1267–1269
15. Olid D, Nunez R, Vinas C et al (2013) Methods to produce B-C, B-P, B-N and B-S bonds in boron clusters. Chem Soc Rev 42(8):3318–3336

Chapter 6
Ene Reaction of 1,3-Dehydro-*o*-Carborane

Abstract This chapter addresses the efficient ene reaction of 1,3-dehydro-*o*-carborane with simple alkenes and alkynes for the synthesis of cage B(3)-alkenyl/allenyl *o*-carboranes, which are not ready to access by known methods. The excellent regioselectivity observed in this reaction can be ascribed to the highly polarized cage C–B multiple bond, which lowers the activation barriers of the ene reaction.

6.1 Introduction

Ene reactions or the Alder ene reaction, which involves the attack of an alkene (the ene donor) by an unsaturated enophile (A=B or A≡B), is well known in organic chemistry [1]. In this process, an allylic hydrogen atom is transferred to the enophile and the alkene π-bond is transposed. Ene donors are π-bonded molecules that contain at least one active hydrogen atom at the allylic, propargylic, or α-position. Possible ene components include olefinic, acetylenic, allenic, aromatic, cyclopropyl, and carbon-heterobonds. Ene donors containing C=O, C=N and C=S bonds have been reported, but such cases are rare [2].

Enophiles are π-bonded molecules that have electron-withdrawing substituents that can lower the LUMO of the π-bond [2]. Particularly, due to the low lying of LUMO level relative to typical alkynes caused by distortion of the bond angles of the alkyne moiety, arynes are highly reactive species that behave as an electrophilic reagent toward a variety of electron-rich species [3]. Ene reaction involving aryne intermediates or the aryne–ene reaction has been developed and widely used in synthesis [4]. In sharp contrast, although ene reaction of *o*-carboryne was well known, such reactions often accompany with [2 + 2] cycloadditions, leading to a poor selectivity and the yield is rather low (Scheme 6.1) [5].

Considering the unique properties of functionalized carboranes witnessed in medicine [6], materials science [7], and organometallic/coordination chemistry [8]

D. Zhao, *Functionalization of Carborane via Carboryne Intermediates*,
Springer Theses, DOI 10.1007/978-981-10-1569-4_6

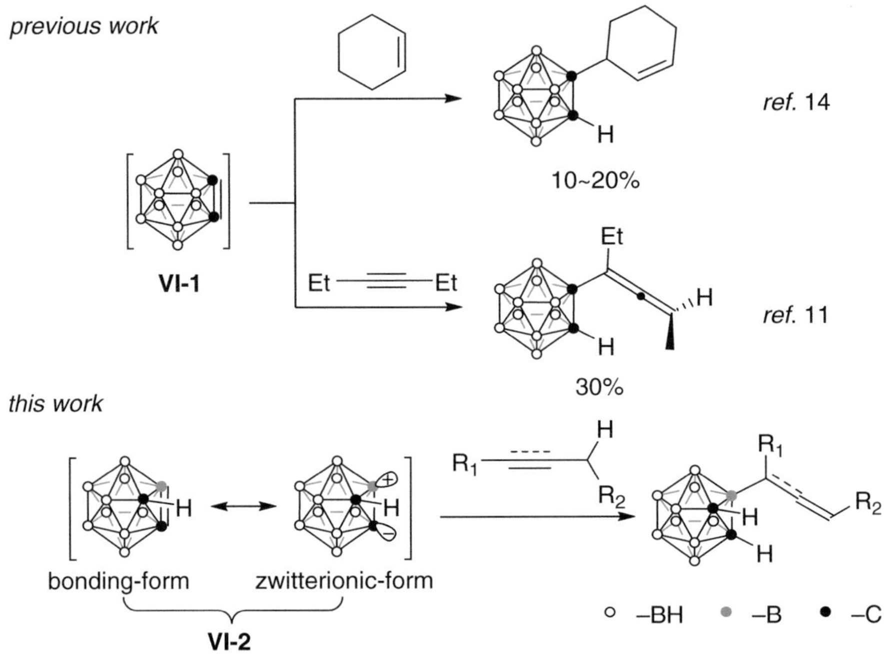

Scheme 6.1 Ene reaction of carboryne intermediates

and the importance of alkenes and allenes as two classes of feedstocks in chemical synthesis, compounds that combine both carborane clusters and allene/alkene can serve as useful starting materials for further functionalization toward various purposes. Unfortunately, straightforward and general synthesis of such cage boron-substituted alkenes/allenes still represents a very challenging task despite recent remarkable progress in carborane chemistry [9, 10]. To the best of our knowledge, protocols to address cage boron-substituted allenes remain elusive in literature. On the other hand, cage carbon-substituted carboranyl allenes [11] can be prepared in low yields (<30 %) from the reaction of 1,2-dehydro-*o*-carborane (*o*-carboryne) (**VI-1**) [12] with alkynes via ene reaction pathway (Scheme 6.1).

An ene reaction can be rationalized as interactions between the HOMO of the ene and LUMO of the enophile [13]. Thus, a more polarized enophile should favor ene reaction. In this regard, our recently reported 1,3-dehydro-*o*-carborane (**VI-2**) [14], featuring a polarized cage C–B multiple bond, would be an excellent candidate for ene reactions with alkenes and alkynes. In this section, we describe a general, high-yielding, and regioselective ene reaction of 1,3-dehydro-*o*-carborane for the straightforward access to cage B(3)-substituted carboranyl alkenes and allenes (Scheme 6.1).

6.2 Ene Reaction of 1,3-Dehydro-*o*-Carborane: Reactivity Toward Alkenes

To test our hypothesis, the reaction of 3-diazonium-*o*-carborane tetrafluoroborate (**VI-3**) with 2,3 dimethyl-2-butene (**VI-4a**) was chosen as a model to establish the optimal reaction conditions (Table 6.1). The results showed that among the non-nucleophilic bases examined, lithium diisopropylamide (LDA) gave the best result, affording the desired ene product **VI-5a** in 92 % isolated yield (Table 6.1, entries 1–4). Increasing the amount of LDA to 2.0 equivalent did not improve the yield (Table 6.1, entry 5). The molecular structure of **VI-5a** was confirmed by single-crystal X-ray analyses (Fig. 6.1). It was noted that no [2 + 2] cycloaddition product was observed in these reactions, suggesting that ene reaction of 1,3-dehydro-*o*-carborane is much more efficient than that of *o*-carboryne [15, 16].

Under the optimal reaction conditions (Table 6.1, entry 1), various alkenes were examined and the results are compiled in Table 6.2. In general, the chemoselectivity is excellent, giving only ene reaction products in very good to excellent yields, in which the substituents are always bonded to cage B(3) position. No cycloaddition species were observed. For 4-methyl-1-pentene **VI-4b** and 2-ethyl-1-butene **VI-4c**, a mixture of *E/Z*-isomers was formed in >87 % yields with a molar ratio of 2.6:1 and 1:1.7, respectively (Table 6.2, entries 2–3). The reaction of methylenecyclohexane **VI-4d** smoothly produced the desired product **VI-5d** in 93 % yield (Table 6.2, entry 4). An evident preference for primary C–H over secondary C–H proton migration was observed for substrate **VI-4e** that contains different allylic hydrogen (Table 6.2, entry 5).

Table 6.1 Optimization of reaction conditions

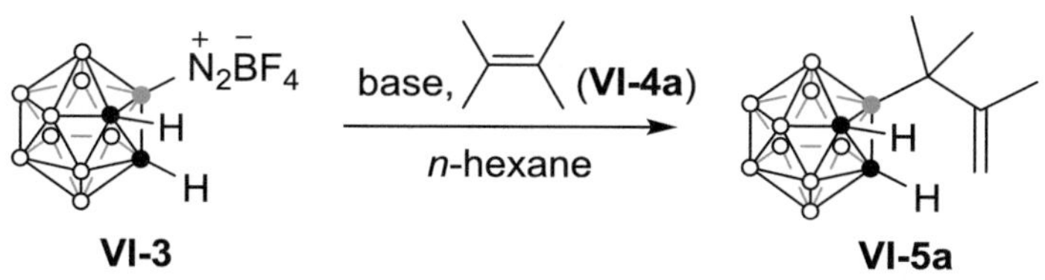

Entry	Base (equiv)	Isolated yields (%)
1	LDA (1.0)	92
2	LiHMDS (1.0)	86
3	NaH (1.0)	63
4	$NaNH_2$ (1.0)	67
5	LDA (2.0)	92

Reaction conditions: **VI-3** (0.2 mmol), base (0.2 or 0.4 mmol), **VI-4a** (1.0 mmol), *n*-hexane (2 mL), −30 °C to rt, 10 min. LDA = lithium diisopropylamide, LiHMDS = lithium bis(trimethylsilyl) amide

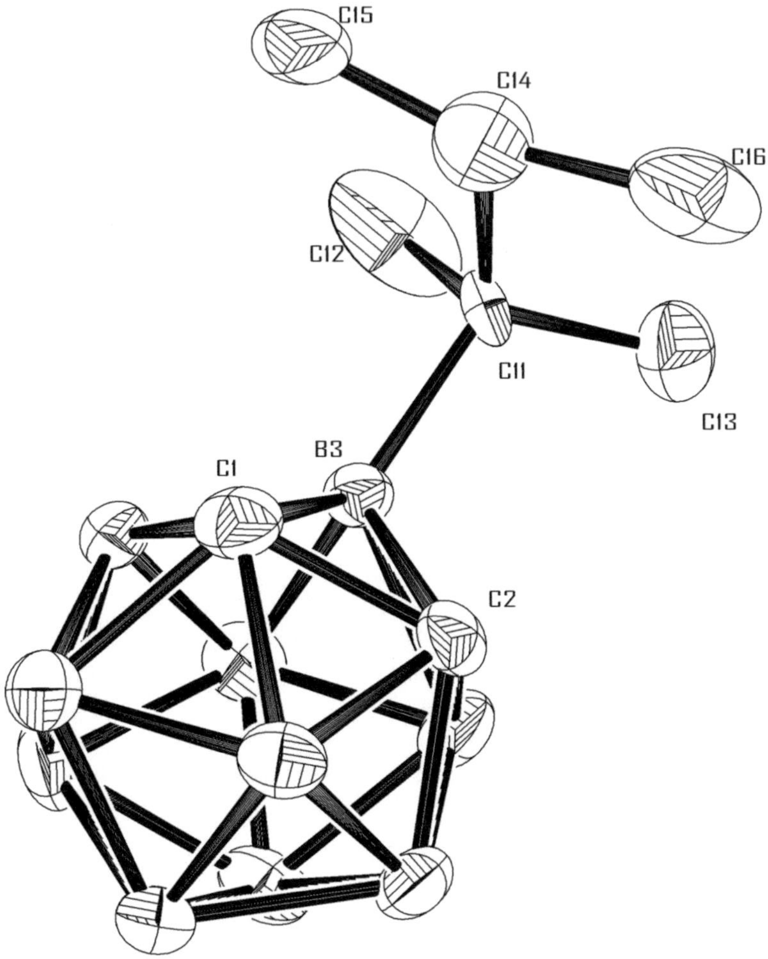

Fig. 6.1 Molecular structure of **VI-5a**

The expected ene reaction products **VI-5f–k** were generated in very good to excellent yields for a series of cyclic olefins (Table 6.2, entries 6–11). Notably, the reactions of 1-piperidino-1-cyclohexene **VI-4h** and 1-morpholino-1-cyclohexene **4i** gave the same product 2-carboranylcyclohexanone **VI-5h** after work-up, which was generated from the hydrolysis of the initially formed ene adduct (Table 6.2, entries 8–9) [17]. For 2,3-dimethyl-1,3-butadiene **VI-4l**, only the ene reaction product **VI-5l** was isolated in 92 % yield (Table 6.2, entry 12). Neither [4 + 2] nor [2 + 2] cycloaddition reaction was observed. The reaction of allene **VI-4m** proceeded smoothly, affording carborane-substituted diene **VI-5m** in 92 % isolated yield (Table 6.2, entry 13).

Table 6.2 Synthesis of B(3)-substituted carboranyl alkenes

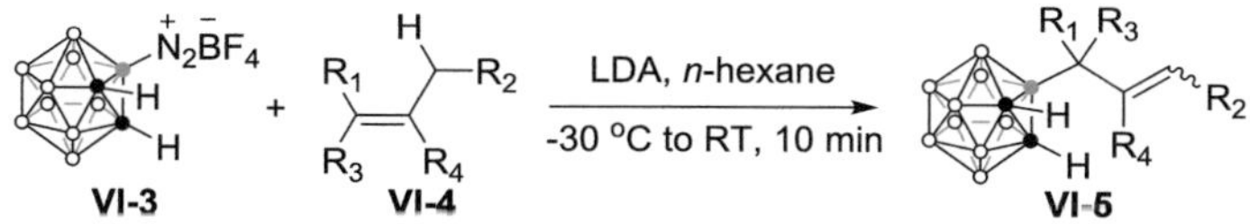

Entry	Alkene **VI-4**	Product **VI-5**[b]	Entry	Alkene **VI-4**	Product **VI-5**[b]
1	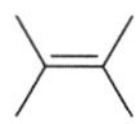 **VI-4a**	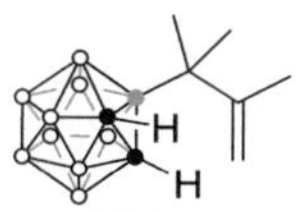 **VI-5a** 92%	7	 **VI-4g**	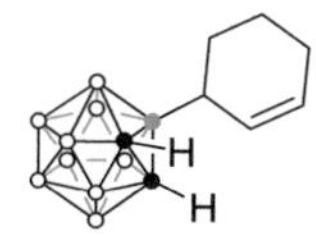 **VI-5g** 91%
2	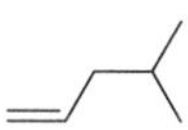 **VI-4b**	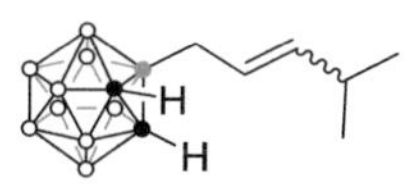 **VI-5b** 87% E/Z = 2.6/1[c]	8,9	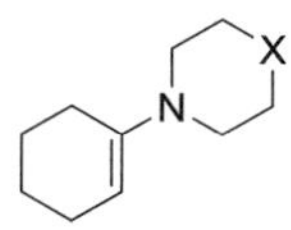 **VI-4h**, X=CH_2 **VI-4i**, X=O	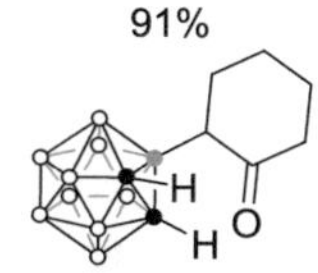 **VI-5h** 98% from **VI-4h** 96% from **VI-4i**
3	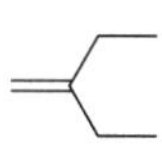 **VI-4c**	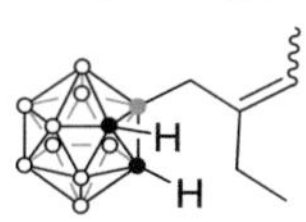 **VI-5c** 97% E/Z = 1/1.7[c]	10	 **VI-4j**	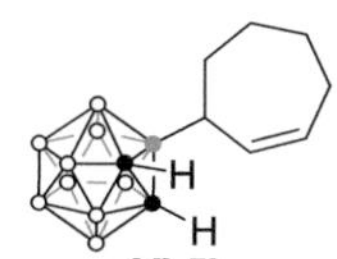 **VI-5j** 83%
4	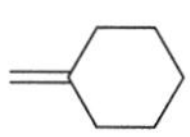 **VI-4d**	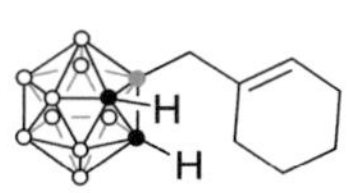 **VI-5d** 93%	11	 **VI-4k**	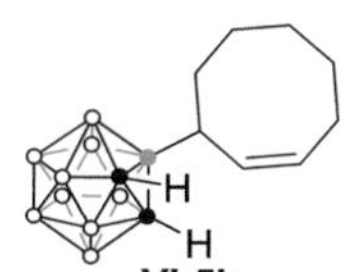 **VI-5k** 76%
5	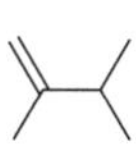 **VI-4e**	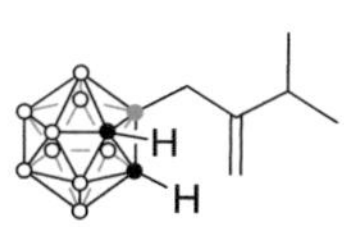 **VI-5ea** 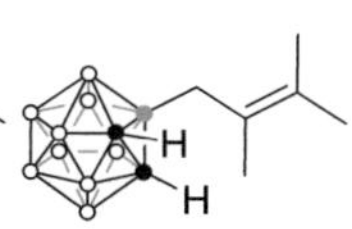 **VI-5eb** 94% **VI-5ea/VI-5eb** = 2:1[c]	12	 **VI-4l**	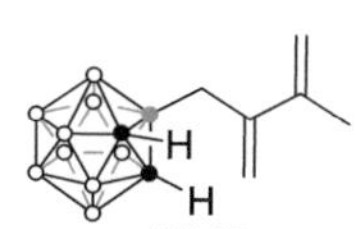 **VI-5l** 92%
6	 **VI-4f**	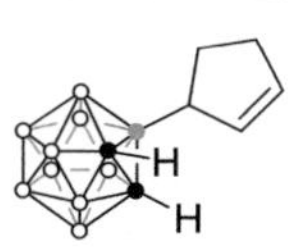 **VI-5f** 83%	13	 **VI-4m**	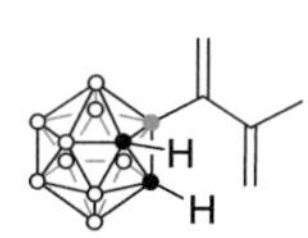 **VI-5m** 92%

[a]Reaction conditions: **VI-3** (0.2 mmol), LDA (0.2 mmol), alkene **VI-4** (1.0 mmol), *n*-hexane (2 mL), −30 °C to rt, 10 min

[b]Isolated yields

[c]Determined by ^{1}H NMR

Table 6.3 Synthesis of B(3)-substituted carboranyl allenes

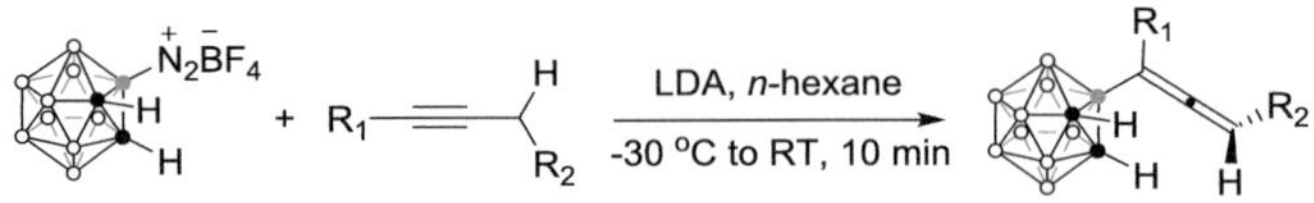

VI-3 **VI-6** **VI-7**

Entry	Alkyne **VI-6**	Product **VI-7**[b]	Entry	Alkyne **VI-6**	Product **VI-7**[b]
1	Et—≡—Et **VI-6a**	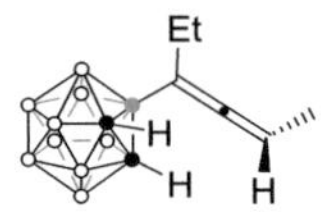 **VI-7a** 92%	7	nPr—≡—TMS **VI-6g**	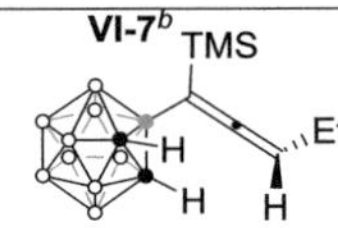 **VI-7g** 83%
2	nPr—≡—nPr **VI-6b**	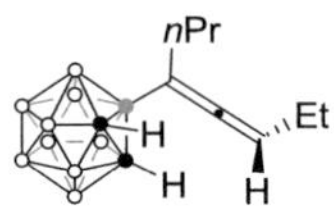 **VI-7b** 90%	8	nBu—≡—TMS **VI-6h**	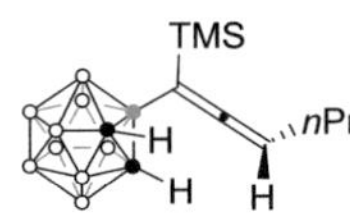 **VI-7h** 78%
3	nBu—≡—nBu **VI-6c**	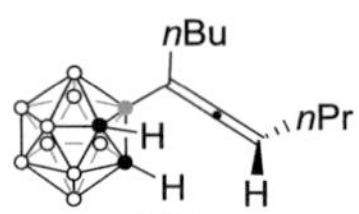 **VI-7c** 89%	9	—≡—tBu **VI-6i**	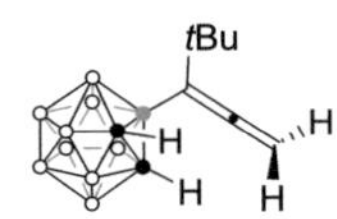 **VI-7i** 72%
4	 **VI-6d**	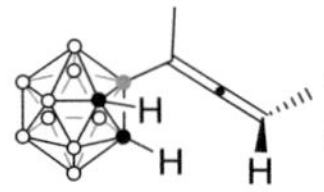 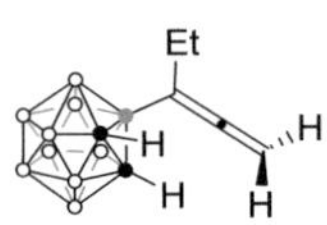 **VI-7da** **VI-7db** 95% **VI-7da/VI-7db** = 1:1.2[c]	10	nPr—≡—≡—nPr **VI-6j**	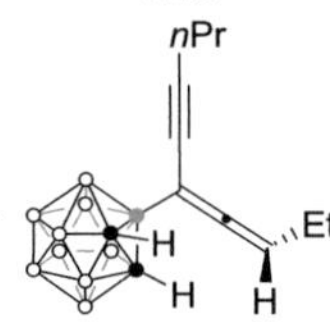 **VI-7j** 85%
5	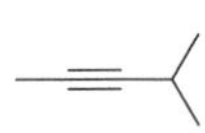 **VI-6e**	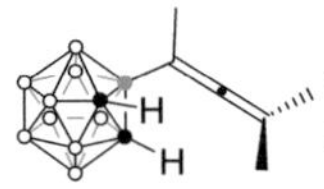 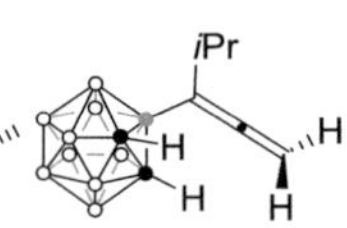 **VI-7ea** **VI-7eb** 94% **VI-7ea/VI-7eb** = 1:2[c]	11	nBu—≡—≡—nBu **VI-6k**	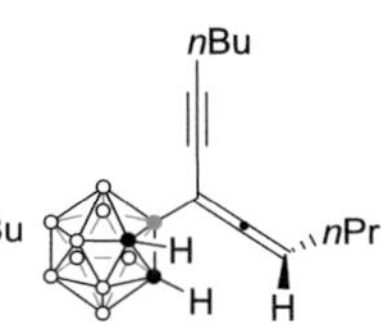 **VI-7k** 80%
6	—≡—TMS **VI-6f**	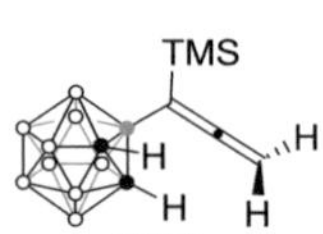 **VI-7f** 87%	12	iPr—≡—≡—iPr **VI-6l**	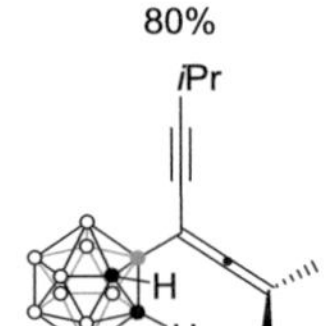 **VI-7l** 78%

[a]Reaction conditions: **VI-3** (0.2 mmol), LDA (0.2 mmol), alkyne **VI-6** (1.0 mmol), n-hexane (2 mL), −30 °C to rt 10 min

[b]Isolated yields

[c]Determined by ^1H NMR

6.3 Ene Reaction of 1,3-Dehydro-*o*-Carborane: Reactivity Toward Alkynes

The aforementioned ene reaction is well compatible with a variety of alkylalkynes, giving cage B(3)-substituted carboranyl allenes in very high to excellent yields. The results are summarized in Table 6.3. Symmetric alkynes **VI-6a–c** worked well to afford the corresponding B(3)-allenyl carboranes in excellent yields (Table 6.3, entries 1–3). As observed in ene reaction of alkene **VI-4e** (Table 6.2, entry 5), a preference for primary C–H over secondary C–H/tertiary C–H proton migration was also found for unsymmetrical alkynes **VI-6d** and **VI-6e** (Table 6.3, entries 4–5). For 1-TMS alkynes **VI-6f–h** and 4,4-dimethyl-2-pentyne **VI-6i**, the ene products were produced in decreased yields, probably due to steric reasons (Table 6.3, entries 6–9). Various symmetrical diynes were also compatible with this reaction, affording the corresponding carborane-substituted alkynyl allenes **VI-7j–l** in high yields (Table 6.3, entries 10–12). This method represents the first simple and efficient approach for B(3)-allenyl-*o*-carboranes.

6.4 Mechanistic Study

To understand the proton transfer process, treatment of deuterium-labeled substrate **VI-4g-d$_{10}$** with **VI-3** gave **VI-5g-d$_{10}$** in 88 % yield with 96 % deuterium incorporation (Eq. 6.1 and Figs. 6.2, 6.3, and 6.4).

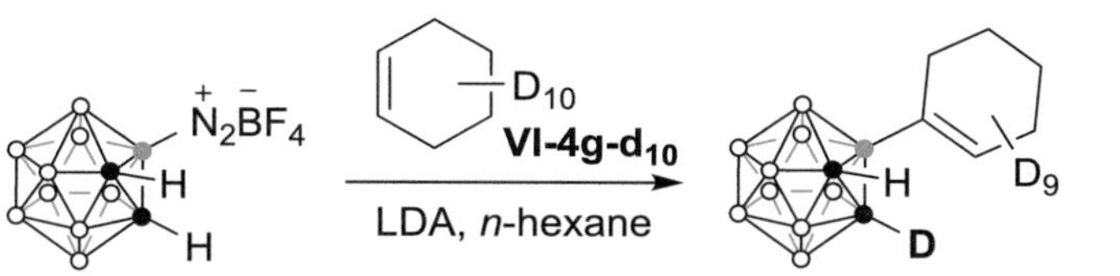

VI-3 **VI-5g-d$_{10}$**, 96%-D, 88% (6.1)

The inherent reactivity differences between alkene and alkyne were evaluated by the reaction of 2-methyl-1-hexen-3-yne (**VI-8**) with **VI-3**. It was found that the ene reaction products **VI-9a** and **VI-9b** were formed in a molar ratio of 1.2/1, indicating the similar reactivity of two reactive sites (Eq. 6.2 and Scheme 6.2).

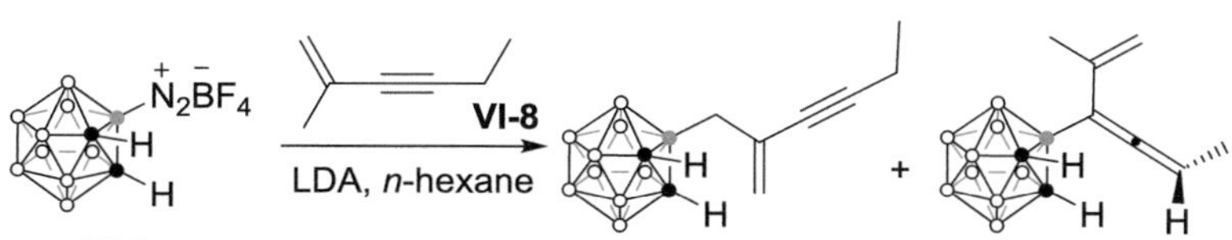

VI-3 **VI-9a** **VI-9b** (6.2)

VI-9a/VI-9b = 1.2/1, 91%

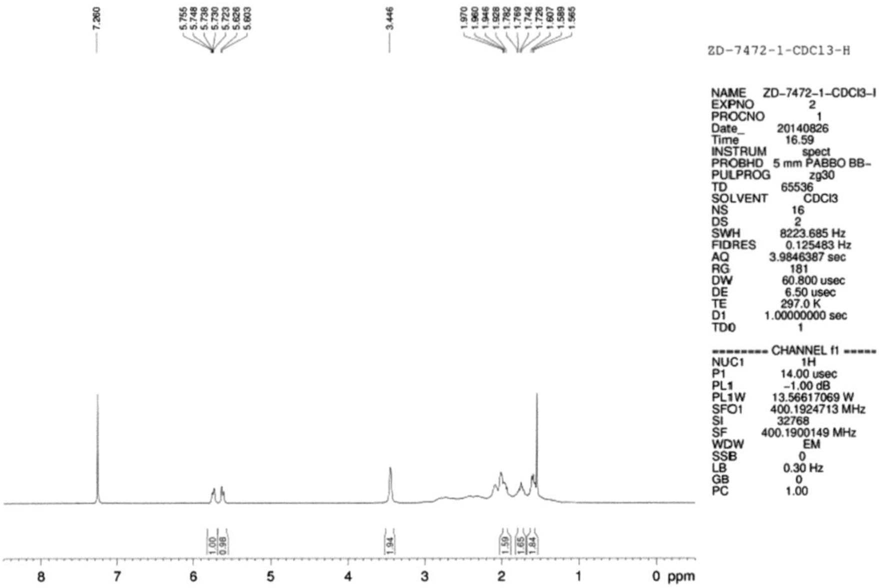

Fig. 6.2 ^{1}H NMR spectrum of **VI-5g** (for reference)

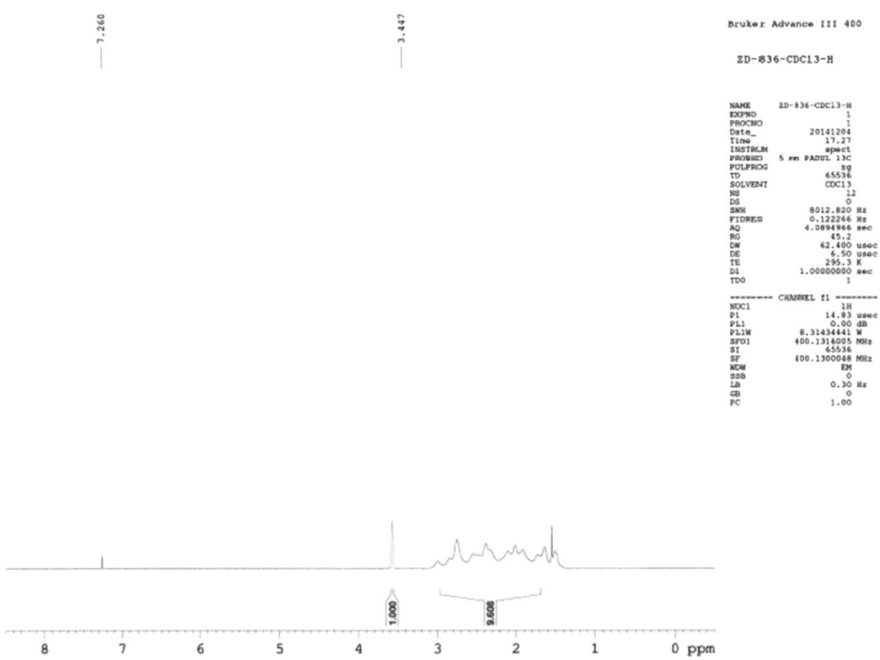

Fig. 6.3 ^{1}H NMR spectrum of **VI-5g-d$_{10}$**

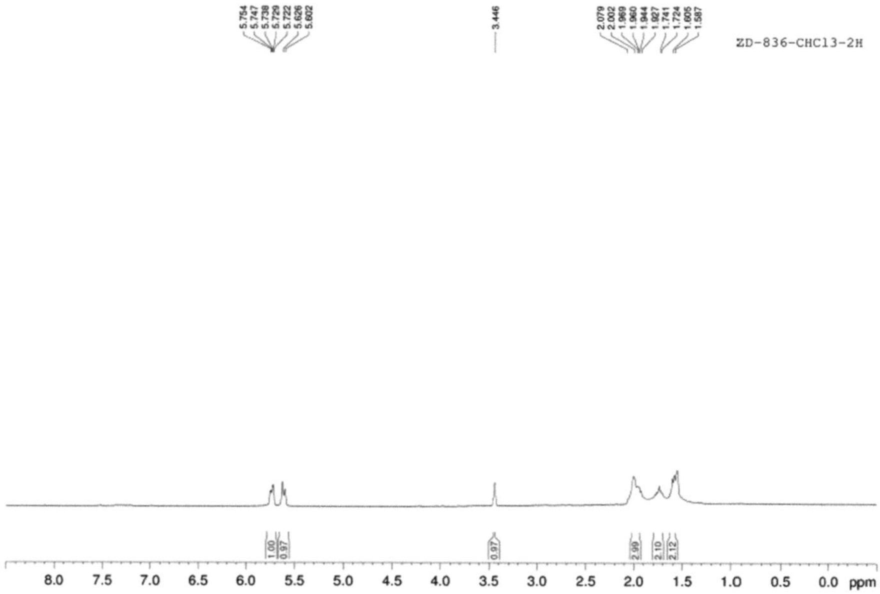

Fig. 6.4 ^{2}H NMR spectrum of **VI-5g-d$_{10}$**

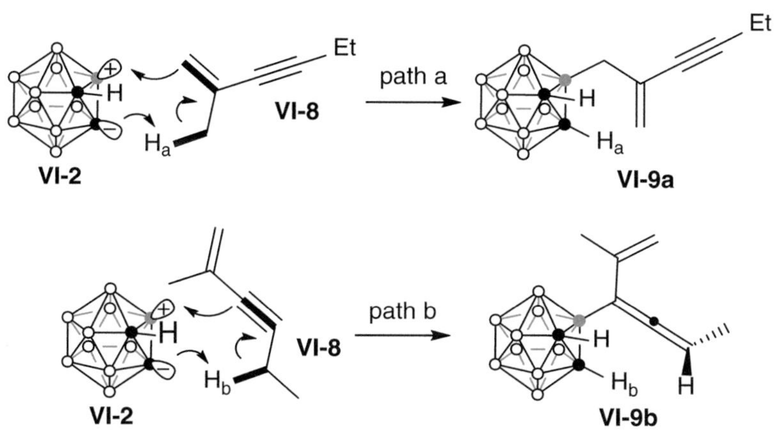

Scheme 6.2 Two possible pathways for 2-methyl-1-hexen-3-yne (**VI-8**)

It was noteworthy that such ene reaction of 1,3-dehydro-*o*-carborane proceeded with excellent regioselectivity, giving only the B(3)-substituted products. This selectivity can be ascribed to the zwitterionic nature of 1,3-dehydro-*o*-carborane as illustrated in Scheme 6.3a [14]. The highly polarized C–B multiple bond can also rationalize that the selectivity of the C–H bonds in the ene reactions follows the

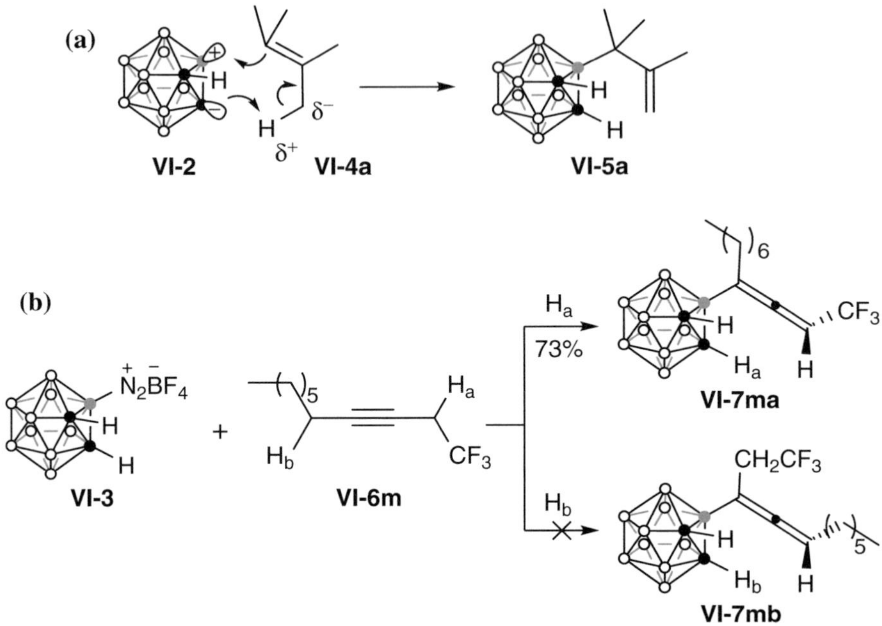

Scheme 6.3 Regioselectivity in ene reaction of 1,3-dehydro-*o*-carborane

order: primary C–H > secondary C–H > tertiary C–H, which is well in agreement with the acidity of these C–H protons [18].

In support of this observation, substrate **VI-6m** processing a CF_3-substituent was subjected to this reaction. Due to the strong inductive effect of the CF_3 group, hydrogen H_a is more acidic than hydrogen H_b. Thus, the migration of H_a should be

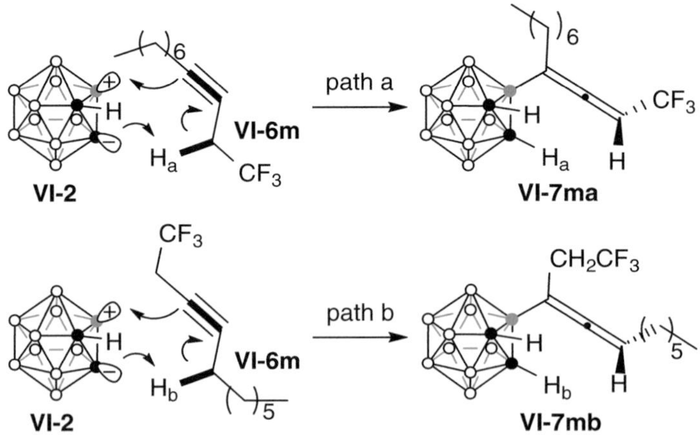

Scheme 6.4 Two possible pathways for 1,1,1-trifluoroundec-3-yne (**VI-6m**)

more favored over that of H_b. In fact, only product **VI-7ma** was isolated in 73 % yield, supporting that the nature of the hydrogen transfer is a proton transfer (Schemes 6.3b and 6.4) [19].

6.5 Summary

1,3-Dehydro-*o*-carborane can readily react with a series of alkenes and alkynes, providing a general, high-yielding synthesis of B(3)-alkenyl/allenyl carboranes with excellent regioselectivity. Ene reactions are dominant in these processes, which can be ascribed to the highly polarized cage C–B multiple bond of 1,3-dehydro-*o*-carborane. The experimental results reveal that the hydrogen shift in the ene reaction is a proton transfer in nature. The present work also demonstrates that 1,3-dehydro-*o*-carborane is a useful synthon for selective cage B-functionalization of *o*-carboranes [20, 21]. The newly prepared carboranyl alkenes and allenes may serve as useful starting materials for further functionalization.

References

1. Hoffmann HMR (1969) The ene reaction. Angew Chem Int Ed Engl 8(8):556–577
2. Paderes GD, Jorgensen WL (1992) Computer-assisted mechanistic evaluation of organic-reactions. 20. Ene and retro-ene chemistry. J Org Chem 57(6):1904–1916
3. (a) Tadross PM, Stoltz BM (2012) A comprehensive history of arynes in natural product total synthesis. Chem Rev 112(6):3550–3577 (b) Pena D, Perez D, Guitian E (2006) Insertion of arynes into sigma bonds. Angew Chem Int Ed 45(22):3579-3581 (c) Wenk HH, Winkler M, Sander W (2003) One century of aryne chemistry. Angew Chem Int Ed 42(5):502–528
4. (a) Jayanth TT, Jeganmohan M, Cheng MJ et al (2006) Ene reaction of arynes with alkynes. J Am Chem Soc 128(7):2232–2233 (b) Aly AA, Mohamed NK, Hassan AA et al (1999) Reaction of diimines and benzyne. Tetrahedron 55(4):1111–1118 (c) Garsky V, Koster DF, Arnold RT (1974) Studies of the stereochemistry and mechanism of the ene reaction using specifically deuterated pinenes. J Am Chem Soc 96(13):4207–4210 (d) Aly AA, Shaker RM (2005) 5-benzyl-1H-tetrazols from the reaction of 1-aryl-5-methyl-1h-tetrazoles with 1,2-dehydrobenzene. Tetrahedron Lett 46(15):2679-2682 (e) Nakayama J, Yoshimura K (1994) A general-synthesis of aromatic-compounds carrying 2 neopentyl groups on adjacent positions. Tetrahedron Lett 35(17):2709-2712 (f) Tabushi I, Okazaki K, Oda R (1969) Relative reactivities of substituted olefins toward benzyne. Tetrahedron 25(18):4401–4407 (g) Crews P, Beard J (1973) Cycloadditions of benzyne with cyclic olefins: competition between 2 + 4, ene, and 2 + 2 reaction pathways. J Org Chem 38(3):522–528
5. (a) Cunningham RJ, Bian NY, Jones M (1994) Reaction of 1,2-dehydro-o-carborane with acetylenes—ene reactions and cycloaddition to give an analog of benzocyclobutadiene. Inorg Chem 33(22):4811–4812 (b) Ho DM, Cunningham RJ, Brewer JA et al (1995) Reaction of 1,2-dehydro-o-carborane with acetylenes—synthesis and structure of a carborane analog of benzocyclobutadiene. Inorg Chem 34(21):5274–5278

6. (a) Kracke GR, VanGordon MR, Sevryugina YV et al (2015) Carborane-derived local anesthetics are isomer dependent. Chemmedchem 10(1):62–67 (b) Wilkinson SM, Gunosewoyo H, Barron ML et al (2014) The first cns-active carborane: a novel P2X7 receptor antagonist with antidepressant activity. Acs Chem Neurosci 5(5):335–339 (c) Scholz M, Kaluderovic GN, Kommera H et al (2011) Carbaboranes as pharmacophores: similarities and differences between aspirin and asborin. Eur J Med Chem 46(4):1131–1139 (d) Beer ML, Lemon J, Valliant JF (2010) Preparation and evaluation of carborane analogues of tamoxifen. J Med Chem 53(22):8012–8020 (e) Julius RL, Farha OK, Chiang J et al (2007) Synthesis and evaluation of transthyretin amyloidosis inhibitors containing carborane pharmacophores. Proc Natl Acad Sci USA 104(12):4808–4813
7. (a) Visbal R, Ospino I, Lopez-de-Luzuriaga JM et al (2013) N-heterocyclic carbene ligands as modulators of luminescence in three-coordinate gold(I) complexes with spectacular quantum yields. J Am Chem Soc 135(12):4712–4715 (b) Cioran AM, Musteti AD, Teixidor F et al (2012) Mercaptocarborane-capped gold nanoparticles: electron pools and ion traps with switchable hydrophilicity. J Am Chem Soc 134(1):212–221 (c) Koshino M, Tanaka T, Solin N et al (2007) Imaging of single organic molecules in motion. Science 316(5826):853–853 (d) Yinghuai Z, Peng AT, Carpenter K et al (2005) Substituted carborane-appended water-soluble single-wall carbon nanotubes: new approach to boron neutron capture therapy drug delivery. J Am Chem Soc 127(27):9875–9880
8. (a) Hosmane NS, Maguire JA (2007) d- and f-Block metallacarbaboranes. In: Crabtree RH, Mingos DMP (eds) Comprehensive organometallic chemistry III. Elsevier, Oxford (b) Xie Z (2002) Advances in the chemistry of metallacarboranes of f-block elements. Coord Chem Rev 231(1–2):23–46 (c) Xie Z (2003) Cyclopentadienyl-carboranyl hybrid compounds: a new class of versatile ligands for organometallic chemistry 36(1):1–9 (d) Deng L, Xie Z (2007) Advances in the chemistry of carboranes and metallacarboranes with more than 12 vertices. Coord Chem Rev 251(17–20):2452–2476 (e) Yao ZJ, Jin GX (2013) Transition metal complexes based on carboranyl ligands containing N, P, and S donors: synthesis, reactivity and applications. Coordin Chem Rev 257(17–18):2522–2535 (f) Zhang J, Xie Z (2014) Synthesis, structure, and reactivity of 13-and 14-vertex carboranes. Acc Chem Res 47(5):1623–1633
9. Grimes RN (2011) Carboranes, 2nd edn. Academic Press, Amsterdam
10. (a) Hosmane NS (2011) Boron science: new technologies and applications. CRC Press, Boca Raton (b) Grimes RN (2015) Carboranes in the chemist's toolbox. Dalton Trans 44(13): 5939–5956 (c) Bregadze VI (1992) Dicarba-closo-dodecaboranes $C_2B_{10}H_{12}$ and their derivatives. Chem Rev 92(2):209–223
11. (a) Zakharkin LI, Koverdov AI, Shaugumbekova ZhS et al (1979) Zhurnal Obshei Khimii 49:1165–1169. (b) Zakharkin LI, Koverdov AI, Shaugumbekova ZhS et al (1981) Zhurnal Obshei Khimii 51:1582–1586
12. (a) Qiu Z, Wang SR, Xie Z (2010) Nickel-catalyzed regioselective [2 + 2+ 2] cycloaddition of carboryne with alkynes. Angew Chem Int Edit 49(27):4649–4652 (b) Qiu Z, Xie Z (2014) Generation and reactivity of o-carborynes. Dalton Trans 43(13):4925–4934
13. Inagaki S, Fujimoto H, Fukui K (1976) Orbital interaction in three systems 98(16):4693–4701
14. Zhao D, Zhang J, Xie Z (2014) 1,3-dehydro-o-carborane: generation and reaction with arenes. Angew Chem Int Ed 53(32):8488–8491
15. Wang SR, Qiu Z, Xie Z (2011) Regioselective insertion of carborynes into ethereal C–H bond: facile synthesis of alpha-carboranylated ethers. J Am Chem Soc 133(15):5760–5763
16. Wang SR, Xie Z (2012) Reaction of carboryne with styrene and its derivatives. Tetrahedron 68 (26):5269–5278
17. Gingrich HL, Huang Q, Morales AL et al (1992) Reactions of enamines with dehydro aromatic compounds. J Org Chem 57(14):3803–3806
18. Bordwell FG (1988) Equilibrium acidities in dimethyl-sulfoxide solution. Accounts Chem Res 21(12):456–463

19. Mikami K, Shimizu M (1992) Asymmetric ene reactions in organic-synthesis. Chem Rev 92(5):1021–1050
20. Olid D, Nunez R, Vinas C et al (2013) Methods to produce B–C, B–P, B–N and B–S bonds in boron clusters. Chem Soc Rev 42(8):3318–3336
21. (a) Qiu Z (2015) Recent advances in transition metal-mediated functionalization of o-carboranes. Tetrahedron Lett 56(8):963–971 (b) Xie Z (2014) Transition metal mediated functionalization of o-carboranes. Sci China Chem 57(8):1061–1063

Chapter 7
Cage Boron Arylation of *o*-Carborane via Metal-Free, Visible-Light-Mediated Radical Coupling

Abstract This chapter demonstrates that 3-diazonium-*o*-carborane tetrafluoroborate can serve as not only an efficient precursor for 1,3-dehydro-*o*-carborane, but also an ideal source of *o*-carboranyl boryl radical in the presence of visible light. For instance, the resulting *o*-carboranyl boryl radical can undergo regioselective C–H insertion reaction with simple arenes bearing a wide range of functional groups. This general and simple procedure provides a transition-metal-free alternative for the synthesis of cage boron-arylated *o*-carboranes, which may found application in materials science. This work also represents a synthetic useful application of boryl radicals.

7.1 Introduction

Direct C–H bond transformation of arenes or heteroarenes is a fundamentally important subject in organic synthesis [1]. Comparing to those metal-catalyzed cross-coupling reactions, the rapidly growing field of photoredox catalysis, which avoids the use of transition metals, ligand, bases, or elevated temperatures, offers a valuable platform for the design of new synthetic transformations [2]. Moreover, the ability of photoredox catalysts to act as both strong oxidants and reductants upon irradiation with low-energy visible light has enabled the invention of a series of useful bond constructions, previously thought to be unattainable via conventional pathways.

On the other hand, due to their potential applications in medicine [3], materials science [4], and organometallic/coordination chemistry [5], there is a growing interest in the design and synthesis of compounds that combine *o*-carborane (icosahedral carboranes or *o*-$C_2B_{10}H_{12}$) clusters and organic units. For instance, carborane derivatives bearing cage π-substituents (aryl, and heteroaryl) have been found interesting photophysical properties as light-emitting materials in recent studies [6].

In sharp contrast, *o*-carborane derivatives with aryl substituents attached to boron have been less investigated compared to their cage C-aryl counterparts [7],

D. Zhao, *Functionalization of Carborane via Carboryne Intermediates*,
Springer Theses, DOI 10.1007/978-981-10-1569-4_7

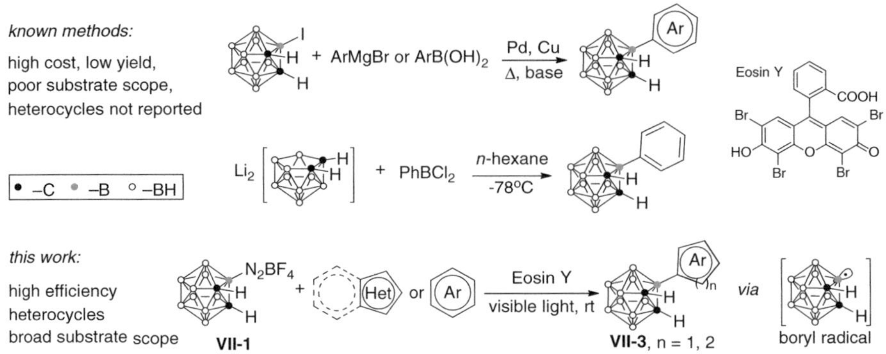

Scheme 7.1 Synthesis of B(3)-aryl carboranes

probably due to the absence of general synthetic methods in the literature [8]. The known methods for the preparation of B-phenyl-*o*-carboranes are either from the Pd-catalyzed cross-coupling reaction of B-I-*o*-carboranes with aryl Grignard reagents/boronic acids [8a, b, e, f, i] or from the capping reaction of $PhBCl_2$ with *nido*-$C_2B_9H_{11}^{2-}$ (Scheme 7.1) [8a, c, d, g, h, j]. Recently, we have reported an Pd-catalyzed intramolecular cross-coupling of B–H bond with aryl bromides [8k]. All these methods always accompany with obvious disadvantages such as low efficiency, high cost, and poor functional group tolerance. Most importantly, carborane derivatives bearing heteroaryl substituents at cage boron still remain elusive, as they cannot be synthesized by the above methods [9]. Thus, it is of great interest and importance to develop an efficient, low-cost methodology for the synthesis of B-(hetero)aryl-*o*-carboranes [10]. In this section, we describe a general, efficient, and regioselective cage boron arylation of *o*-carborane via photoredox catalysis, which avoids the use of any metals, additives, or high temperatures.

7.2 B(3)-Arylation of *o*-Carborane: A Radical Approach

We have recently reported the synthesis of 3-diazonium-*o*-carborane tetrafluoroborate (**VII-1**) for the generation of 1,3-dehydro-*o*-carborane [11]. Inspired by the chemistry of phenyl diazonium salt [12], we wonder whether the carboranyl boryl radical would be generated from **VII-1** through a single-electron transfer (SET) process in the presence of photocatalyst and visible light [13]. Considering the high reactivity of boryl radicals [14], the cage boron aryl-substituted *o*-carboranes may be produced in the presence of arenes or heteroarenes as trapping agents (Scheme 7.1).

To test our hypothesis, direct carboranylation of thiophene (**VII-2a**) with diazonium salt **VII-1** in the presence of 2 % eosin Y as photoredox catalyst, a xenon lamp (optical power of 50 W) as light source, was conducted to establish the

Table 7.1 Optimization of reaction conditions[a]

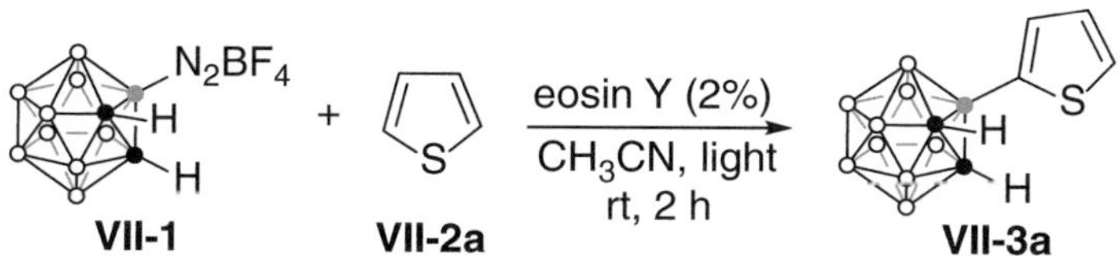

Entry	Conditions	VII-3a[b]
1	**VII-2a** (10 equiv)	83
2	**VII-2a** (5 equiv)	89
3	**VII-2a** (2 equiv)	86
4[c]	**VII-2a** (5 equiv), no catalyst, no light, $Pd(OAc)_2$ (10 %)	Trace
5[c]	**VII-2a** (5 equiv), no catalyst, 12 h	35
6[d]	**VII-2a** (5 equiv), no light, 12 h	Trace

[a]Reaction conditions: **VII-1** (0.1 mmol), **VII-2a** (0.2, 0.5 or 1.0 mmol), eosin Y (2 mol %), CH_3CN (1 mL)
[b]Isolated yields
[c]Main products are 3-F-*o*-carborane and 3-$NHCOCH_3$-*o*-carborane
[d]Main products are starting material **VII-1**

optimal conditions (Table 7.1). The carborane-substituted thiophene **VII-3a** was obtained in very good yields in the presence of 2 equiv of **VII-2a** (Table 7.1, entries 1–3). Only trace amount of desired product was formed when palladium was used as catalysis (Table 7.1, entry 4). Control experiments confirmed that both light and eosin Y are required for significant conversion to the product (Table 7.1, entries 5–6). It is noted that this reaction proceeded with excellent regioselectivity as only substitution at the C2 position of the thiophene was observed. The molecular structure of **VII-3a** is shown in Fig. 7.1.

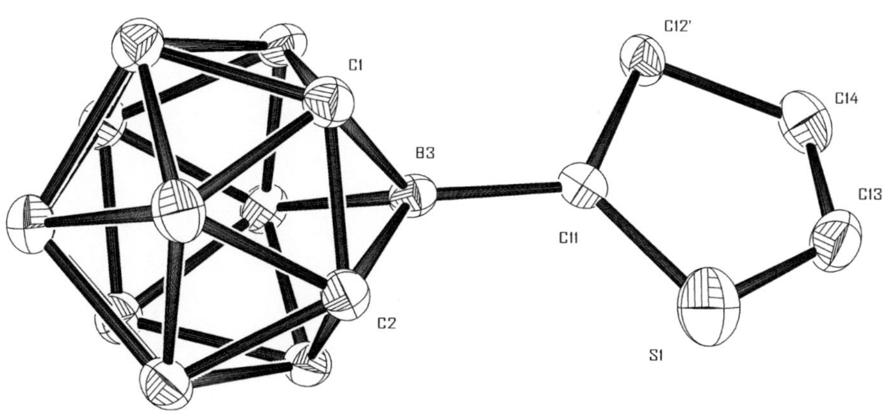

Fig. 7.1 Molecular structure of **VII-3a**

Table 7.2 Scope of heteroarenes[a]

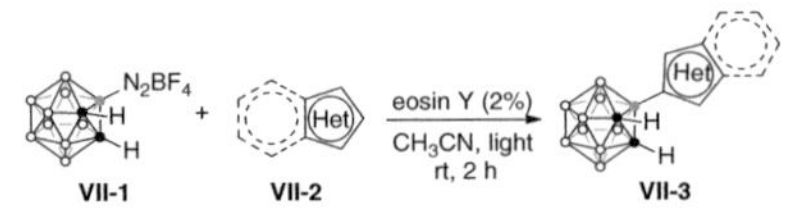

VII-1 VII-2 VII-3

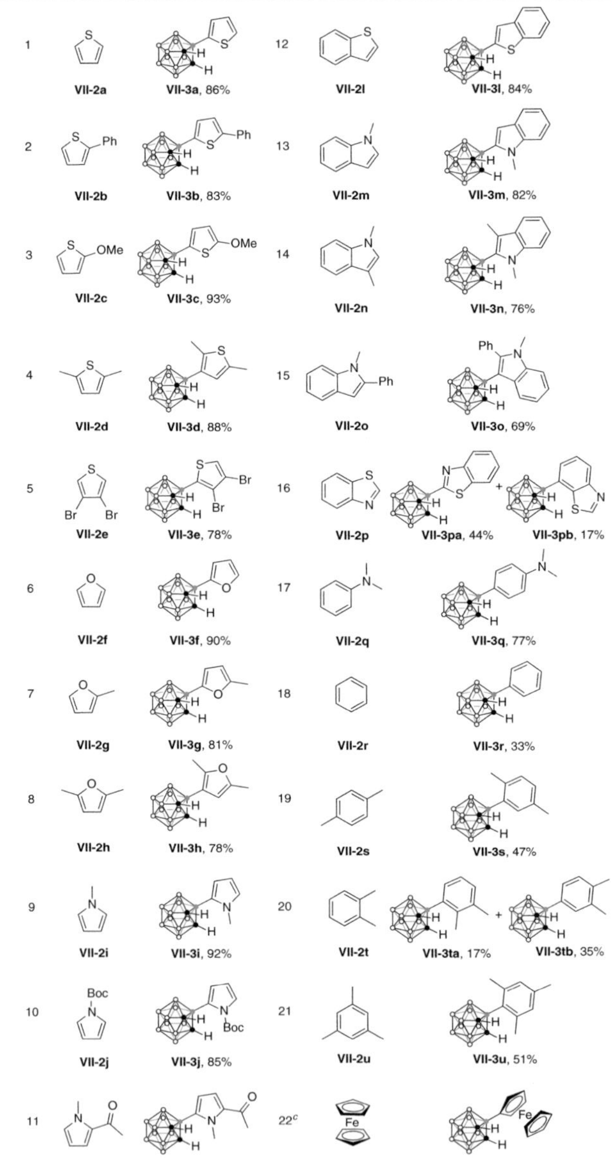

entry	VII-2	VII-3[b]	entry	VII-2	VII-3[b]
1	**VII-2a**	**VII-3a**, 86%	12	**VII-2l**	**VII-3l**, 84%
2	**VII-2b**	**VII-3b**, 83%	13	**VII-2m**	**VII-3m**, 82%
3	**VII-2c**	**VII-3c**, 93%	14	**VII-2n**	**VII-3n**, 76%
4	**VII-2d**	**VII-3d**, 88%	15	**VII-2o**	**VII-3o**, 69%
5	**VII-2e**	**VII-3e**, 78%	16	**VII-2p**	**VII-3pa**, 44% + **VII-3pb**, 17%
6	**VII-2f**	**VII-3f**, 90%	17	**VII-2q**	**VII-3q**, 77%
7	**VII-2g**	**VII-3g**, 81%	18	**VII-2r**	**VII-3r**, 33%
8	**VII-2h**	**VII-3h**, 78%	19	**VII-2s**	**VII-3s**, 47%
9	**VII-2i**	**VII-3i**, 92%	20	**VII-2t**	**VII-3ta**, 17% + **VII-3tb**, 35%
10	**VII-2j**	**VII-3j**, 85%	21	**VII-2u**	**VII-3u**, 51%
11	**VII-2k**	**VII-3k**, 81%	22[c]	**VII-2v**	**VII-3v**, quant

[a]Reaction conditions: **VII-1** (0.1 mmol), heteroarene **VII-2a** (0.5 mmol in the case of thiophene derivatives and 0.2 mmol in the case of other heteroarene), eosin (2 mol %), CH_3CN (1 mL)

[b]Isolated yields

[c]1.0 equiv of ferrocene was used, 5 min

Under the optimal reaction conditions (Table 7.1, entry 3), various thiophene substrates can undergo C–H carboranylation with **VII-1** (Table 7.2, entries 1–5). Such methodology was also compatible to a variety of heterocycles in addition to thiophenes. Various substituents such as alkyl, halogen, alkoxy, carbonyl, and phenyl groups were well tolerated and the desired arylated products **VII-3** were produced in very good to excellent yields. It is noted that this reaction proceeded with excellent regioselectivity as C2 carboranylation was dominant in most cases. Only if the C2 position was unavailable, substitution at C3 position would take place instead (Table 7.2, entries 4, 8, 15). For benzothiazole **VII-2s**, a mixture of C2- and C7-carboranylated products was formed with a molar ratio of 5:2 (Table 7.2, entry 16).

It is noteworthy that for **VII-2b** and **VII-2o**, which feature a phenyl substituent, substitution at the phenyl ring was not observed (Table 7.2, entries 2 and 15). These results may indicate that the electron-rich arenes are more favored compared to the less electron-rich ones, which can also be anticipated from the electrophilic nature of boryl radicals [14]. To further explore the efficiency of this light-mediated coupling reaction, less electron-rich arenes were also examined (Table 7.2, entries 16–21). In general, comparing to those electron-rich ones, the efficiency of these arenes was relative lower as 10 equiv of substrate was needed for a high conversion. This B(3)-arylation process could also be extended to simple benzene derivatives **VII-2q–u**, affording the desired B(3)-aryl carborane **VII-3q–u** in 38–75 % isolated yields (Table 7.2, entries 17–21) [11]. Notably, for dimethyl aniline **VII-2q**, the C–H carboranylation occurred only at *para* position (Table 7.2, entry 17). Moreover, for ferrocene **VII-2m**, which can also serve as a one-electron reductant, the desired product [15] was obtained quantitatively in 5 min with only one equiv of ferrocene (Table 7.2, entry 22). The molecular structures of **VII-3e** and **VII-3v** are shown in Figs. 7.2 and 7.3.

7.3 Mechanistic Study

The C–H carboranylation of heteroarenes with **VII-1** using eosin Y was expected to proceed through a radical mechanism [13], and preliminary mechanistic investigations supported this assumption. Addition of 2,2,6,6-tetramethylpiperidinyloxy (TEMPO) to the reaction mixture completely stopped this carboranylation process and the TEMPO-trapped intermediated **VII-4** was isolated and characterized, suggesting that a radical pathway is involved in this photoreaction (Eq. 7.1)

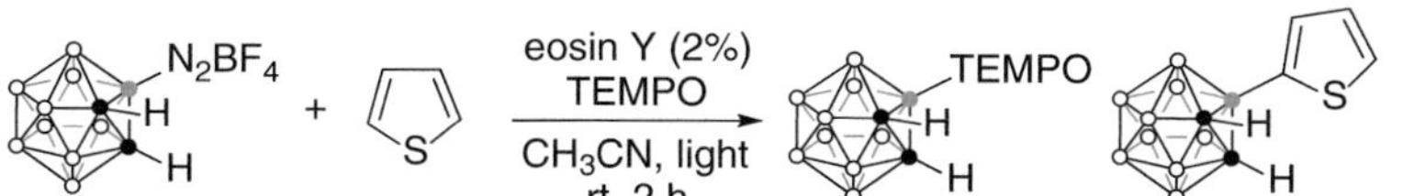

VII-1 **VII-2a** **VII-4**, 76% **VII-3a**, not formed (7.1)

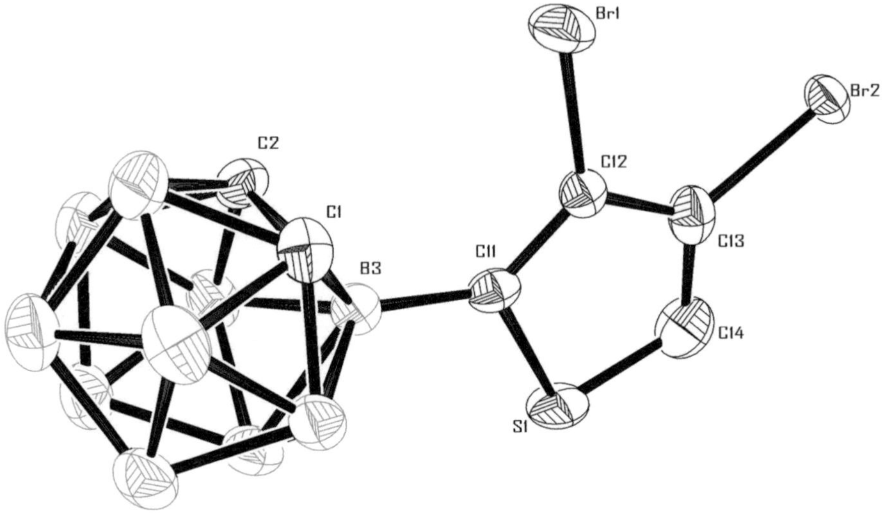

Fig. 7.2 Molecular structure of **VII-3e**

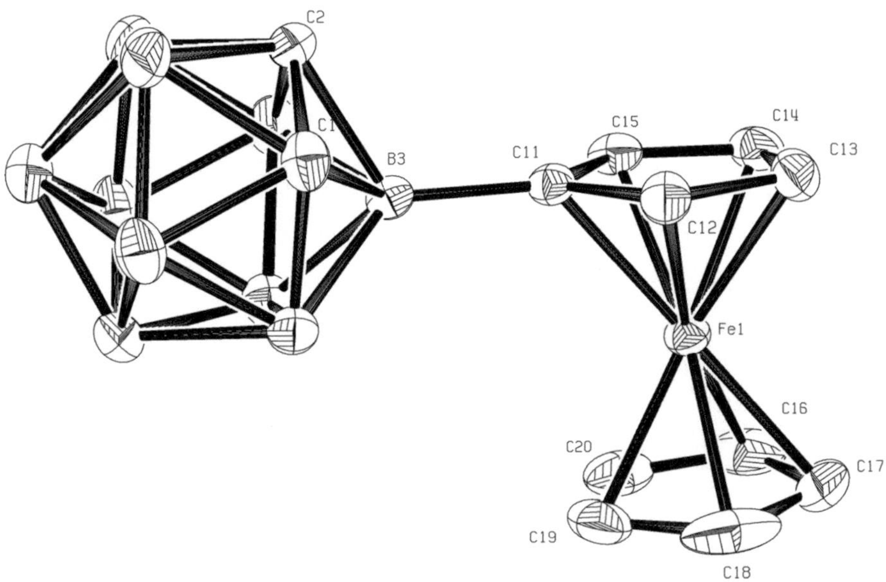

Fig. 7.3 Molecular structure of **VII-3v**

On the basis of the above experimental results, a plausible reaction mechanism is proposed in Scheme 7.2. Irradiation of Eosin Y by visible light generates a photoexcited species, eosin Y*, which could undergo single-electron transfer (SET) with **VII-1** to afford a carboranyl boryl radical **VII-A**. Addition of radical

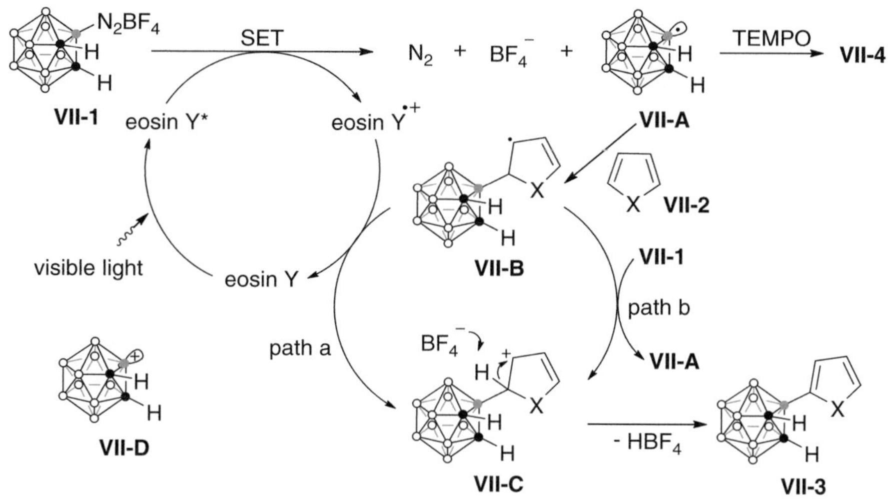

Scheme 7.2 Proposed reaction pathways

VII-A to arene **VII-2** gives an intermediate **VII-B**. Oxidation of intermediate **VII-B** by Eosin Y radical cation (path a) or diazonium salt **VII-1** through a radical chain transfer (path b) would generate a carbocation intermediate **VII-C**. Deprotonation of intermediate **VII-C** leads to the final rearomatized product **VII-3**.

Another possibility that **VII-3** is produced from the Friedel–Crafts reaction involving carboranyl boronium cation **VII-D** [16] can be ruled out by the following observations. First, for indole **VII-2m** that can easily undergo Friedel–Crafts reaction at C3 position, carboranylation occurred only at the C2 position (Table 7.2, entry 13). This is consistent with the fact that radicals preferentially tend to add at C2 position of the indole nucleus. Secondly, if carboranyl boronium cation **VII-D** was formed in the reaction, it would be trapped by the solvent CH_3CN [17]. However, no such adduct was observed.

7.4 Summary

We have developed a metal-free, light-mediated method for selective B(3)-arylation of carborane. The reaction proceeds smoothly at room temperature and simple arenes with a wide range of functional groups can be utilized as coupling partners. This methodology provides a simple and efficient approach to B(3)-aryl carboranes, which may found application in materials science. This work also represents a synthetic useful application of boryl radicals.

References

1. (a) Jones W, Feher, F (1989) Comparative reactivities of hydrocarbon carbon-hydrogen bonds with a transition-metal complex. Acc Chem Res 22(3):91–100 (b) Labinger JA, Bercaw JE (2002) Understanding and exploiting C–H bond activation. Nature 417(6888):507–514 (c) Dyker G (2005) Handbook of C–H transformations: Applications in organic synthesis. Wiley-VCH, Weinheim (d) Godula K, Sames D (2006) C–H bond functionalization in complex organic synthesis. Science 312(5770):67–72 (e) Bergman RG (2007) Organometallic chemistry: C–H activation. Nature 446(7134):391–393
2. (a) Narayanam JMR, Stephenson CRJ (2011) Visible light photoredox catalysis: Applications in organic synthesis. Chem Soc Rev 40(1):102–113 (b) Prier CK, Rankic DA, MacMillan DWC (2013) Visible light photoredox catalysis with transition metal complexes: Applications in organic synthesis. Chem Rev 113(7):5322–5363 (c) Schultz DM, Yoon TP (2014) Solar synthesis: Prospects in visible light photocatalysis. Science 343(6174):985
3. (a) Kracke GR, VanGordon MR, Sevryugina YV et al (2015) Carborane-derived local anesthetics are isomer dependent. Chemmedchem 10(1):62–67 (b) Wilkinson SM, Gunosewoyo H, Barron ML et al (2014) The first cns-active carborane: A novel P2X7 receptor antagonist with antidepressant activity. Acs Chem Neurosci 5(5):335–339 (c) Scholz M, Kaluderovic GN, Kommera H et al (2011) Carbaboranes as pharmacophores: Similarities and differences between aspirin and asborin. Eur J Med Chem 46(4):1131–1139 (d) Beer ML, Lemon J, Valliant JF (2010) Preparation and evaluation of carborane analogues of tamoxifen. J Med Chem 53(22):8012–8020 (e) Julius RL, Farha OK, Chiang J et al (2007) Synthesis and evaluation of transthyretin amyloidosis inhibitors containing carborane pharmacophores. Proc Natl Acad Sci USA 104(12):4808–4813
4. (a) Visbal R, Ospino I, Lopez-de-Luzuriaga JM et al (2013) N-heterocyclic carbene ligands as modulators of luminescence in three-coordinate gold (I) complexes with spectacular quantum yields. J Am Chem Soc 135(12):4712–4715 (b) Cioran AM, Musteti AD, Teixidor F et al (2012) Mercaptocarborane-capped gold nanoparticles: Electron pools and ion traps with switchable hydrophilicity. J Am Chem Soc 134(1):212–221 (c) Koshino M, Tanaka T, Solin N et al (2007) Imaging of single organic molecules in motion. Science 316(5826):853–853 (d) Yinghuai Z, Peng AT, Carpenter K et al (2005) Substituted carborane-appended water-soluble single-wall carbon nanotubes: New approach to boron neutron capture therapy drug delivery. J Am Chem Soc 127(27):9875–9880
5. (a) Hosmane NS, Maguire JA (2007) d- and f-Block metallacarbaboranes in Crabtree RH, Mingos DMP (eds) Comprehensive Organometallic Chemistry III, Elsevier, Oxford (b) Xie Z (2002) Advances in the chemistry of metallacarboranes of f-block elements. Coord Chem Rev 231(1–2):23–46 (c) Xie Z (2003) Cyclopentadienyl-carboranyl hybrid compounds: a new class of versatile ligands for organometallic chemistry. 36(1):1–9 (d) Deng L, Xie Z (2007) Advances in the chemistry of carboranes and metallacarboranes with more than 12 vertices. Coord Chem Rev 251(17–20):2452–2476 (e) Yao ZJ, Jin GX (2013) Transition metal complexes based on carboranyl ligands containing N, P, and S donors: Synthesis, reactivity and applications. Coordin Chem Rev 257(17–18):2522–2535 (f) Zhang J, Xie Z (2014) Synthesis, structure, and reactivity of 13- and 14-vertex carboranes. Acc Chem Res 47 (5):1623–1633
6. (a) Wee KR, Han WS, Cho DW et al (2012) Carborane photochemistry triggered by aryl substitution: Carborane-based dyads with phenyl carbazoles. Angew Chem Int Ed 51 (11):2677–2680 (b) Wee KR, Cho YJ, Song JK et al (2013) Multiple photoluminescence from 1,2-dinaphthyl-ortho-carborane. Angew Chem Int Ed 52(37):9682–9685 (c) Wee KR, Cho YJ, Jeong S et al (2012) Carborane-based optoelectronically active organic molecules: Wide band gap host materials for blue phosphorescence. J Am Chem Soc 134(43):17982–17990
7. (a) Wee KR, Han WS, Cho DW et al (2012) Carborane photochemistry triggered by aryl substitution: Carborane-based dyads with phenyl carbazoles. Angew Chem Int Ed 51 (11):2677–2680 (b) Wee KR, Cho YJ, Jeong S et al (2012) Carborane-based optoelectronically

active organic molecules: Wide band gap host materials for blue phosphorescence. J Am Chem Soc 134(43):17982–17990 (c) Ferrer-Ugalde A, Juarez-Perez EJ, Teixidor F et al (2012) Synthesis and characterization of new fluorescent styrene-containing carborane derivatives: The singular quenching role of a phenyl substituent. Chem-Eur J 18(2):544–553 (d) Wee KR, Cho YJ, Song JK et al (2013) Multiple photoluminescence from 1,2-dinaphthyl-ortho-carborane. Angew Chem Int Ed 52(37):9682–9685 (e) Shi C, Sun HB, Tang X et al (2013) Variable photophysical properties of phosphorescent iridium(iii) complexes triggered by closo- and nido-carborane substitution. Angew Chem Int Ed 52(50):13434–13438 (f) Shi C, Sun HB, Jiang QB et al (2013) Carborane tuning of photophysical properties of phosphorescent iridium (iii) complexes. Chem Commun 49 (42):4746–4748 (g) Bae HJ, Chung J, Kim H et al (2014) Deep red phosphorescence of cyclometalated iridium complexes by *o*-carborane substitution. Inorg Chem 53(1):128–138 (h) Naito H, Morisaki Y, Chujo Y (2015) *o*-Carborane-based anthracene: A variety of emission behaviors. Angew Chem Int Ed 54 (17):5084–5087 (i) Ferrer-Ugalde A, Gonzalez-Campo A, Vinas C et al (2014) Fluorescence of new *o*-carborane compounds with different fluorophores: Can it be tuned? Chem-Eur J 20 (32):9940–9951 (j) Tang C, Xie Z (2015) Nickel-catalyzed cross-coupling reactions of o-carboranyl with aryl iodides: Facile synthesis of 1-aryl-o-carboranes and 1,2-diaryl-o-carboranes. Angew Chem Int Ed 54(26):7662–7665

8. (a) Vinas C, Barbera G, Oliva JM et al (2001) Are halocarboranes suitable for substitution reactions? The case for 3-I-1,2-closo- $C_2B_{10}H_{11}$: Molecular orbital calculations, aryldehalogenation reactions, ^{11}B NMR interpretation of closo-carboranes, and molecular structures of 1-Ph-3-Br-1,2-closo- $C_2B_{10}H_{10}$ and 3-Ph-1,2-closo-$C_2B_{10}H_{11}$. Inorg Chem 40 (26):6555–6562 (b) Aizawa K, Ohta K, Endo Y (2010) Synthesis of 3-aryl-1,2-dicarba-closo-dodecaboranes by suzuki-miyaura coupling reaction. Heterocycles 80(1):369–377 (c) Jin GF, Hwang JH, Lee JD et al (2013) A three-dimensional pi-electron acceptor, tri-phenyl-o-carborane, bearing a rigid conformation with end-on phenyl units. Chem Commun 49 (82):9398–9400 (d) Barbera G, Vaca A, Teixidor F et al (2008) Designed synthesis of new ortho-carborane derivatives: From mono- to polysubstituted frameworks. Inorg Chem 47(16):7309–7316 (e) Ohta K, Yamazaki H, Endo Y (2009) Magnesium-assisted intramolecular demethylation utilizing carborane C-H geometry. J Organomet Chem 694 (11):1646–1651 (f) Eriksson L, Beletskaya IP, Bregadze VI et al (2002) Palladium-catalyzed cross-coupling reactions of arylboronic acids and 2-i-p-carborane. J Organomet Chem 657(1–2):267–272 (g) Hawthorn.Mf, Wegner PA (1968) Reconstruction of 1,2-dicarbaclovododeca-borane(12) structure by boron-atom insertion with (3)-1,2-dicarbollide ions. J Am Chem Soc 90(4):896–901. (h) Roscoe JS, Kongpric.S, Papetti S (1970) Icosahedral carboranes .14. Preparation of boron-substituted carboranes by boron-insertion reaction. Inorg Chem 9 (6):1561–1563. (i) Ogawa T, Ohta K, Yoshimi T et al (2006) M-carborane bisphenol structure as a pharmacophore for selective estrogen receptor modulators. Bioorg Med Chem Lett 16 (15):3943–3946 (j) Safronov AV, Shlyakhtina NI, Hawthorne MF (2012) New approach to the synthesis of 3-alkyl-1,2-dicarba-closo-dodecaboranes: Reaction of alkyldichloroboranes with thallium dicarbollide. Organometallics 31(7):2764–2769 (k) Quan YJ, Xie ZW (2015) Palladium-catalyzed regioselective intramolecular coupling of *o*-carborane with aromatics via direct cage b-h activation. J Am Chem Soc 137(10):3502–3505 (l) Cao K, Huang YW, Yang JX et al (2015) Palladium catalyzed selective mono-arylation of *o*-carboranes via B-H activation. Chem Commun 51(33):7257–7260
9. Beletskaya IP, Bregadze VI, Kabytaev KZ et al (2007) Palladium-catalyzed amination of 2-iodo-para-carborane. Organometallics 26(9):2340–2347
10. Vondrak T, Plesek J, Hermanek S et al (1989) Charge-distribution in icosahedral carboranes: a UV photoelectron spectroscopic study. Polyhedron 8(6):805–811
11. Zhao D, Zhang J, Xie Z (2014) 1,3-dehydro-o-carborane: Generation and reaction with arenes. Angew Chem Int Ed 53(32):8488–8491
12. (a) Mo FY, Dong GB, Zhang Y et al (2013) Recent applications of arene diazonium salts in organic synthesis. Org Biomol Chem 11(10):1582–1593 (b) Hari DP, Konig B (2013) The

photocatalyzed meerwein arylation: Classic reaction of aryl diazonium salts in a new light. Angew Chem Int Edit 52(18):4734–4743 (c) Ando, W (2010) Diazonium and Diazo Groups. Wiley, Chichester (d) Mahouche-Chergui S, Gam-Derouich S, Mangeney C et al (2011) Aryl diazonium salts: A new class of coupling agents for bonding polymers, biomacromolecules and nanoparticles to surfaces. Chem Soc Rev 40(7):4143–4166 (e) Roglans A, Pla-Quintana A, Moreno-Manas M (2006) Diazonium salts as substrates in palladium-catalyzed cross-coupling reactions. Chem Rev 106(11):4622–4643 (f) Felpin FX, Nassar-Hardy L, Le Callonnec F et al (2011) Recent advances in the heck-matsuda reaction in heterocyclic chemistry. Tetrahedron 67(16):2815–2831 (g) Taylor JG, Moro AV, Correia CRD (2011) Evolution and synthetic applications of the heck-matsuda reaction: The return of arenediazonium salts to prominence. Eur J Org Chem (8):1403–1428

13. (a) Kalyani D, McMurtrey KB, Neufeldt SR et al (2011) Room-temperature C-H arylation: Merger of pd-catalyzed c-h functionalization and visible-light photocatalysis. J Am Chem Soc 133(46):18566–18569 (b) Hari DP, Schroll P, Konig B (2012) Metal-free, visible-light-mediated direct C-H arylation of heteroarenes with aryl diazonium salts. J Am Chem Soc 134(6):2958–2961 (c) Heinrich MR, Wetzel A, Kirschstein M (2007) Intermolecular radical carboaminohydroxylation of olefins with aryl diazonium salts and tempo. Org Lett 9 (19):3833–3835 (d) Pratsch G, Anger CA, Ritter K et al (2011) Hydroxy- and aminophenyl radicals from arenediazonium salts. Chem-Eur J 17(15):4104–4108 (e) Wetzel A, Ehrhardt V, Heinrich MR (2008) Synthesis of amino- and hydroxybiphenyls by radical chain reaction of arenediazonium salts. Angew Chem Int Ed 47(47):9130–9133 (f) Wetzel A, Pratsch G, Kolb R et al (2010) Radical arylation of phenols, phenyl ethers, and furans. Chem-Eur J 16(8):2547–2556 (g) Lazarides T, McCormick T, Du PW et al (2009) Making hydrogen from water using a homogeneous system without noble metals. J Am Chem Soc 131(26):9192–9194 (h) Liu HJ, Feng W, Kee CW et al (2010) Organic dye photocatalyzed alpha-oxyamination through irradiation with visible light. Green Chem 12 (6):953–956 (i) ari DP, Hering T, Konig B (2012) Visible light photocatalytic synthesis of benzothiophenes. Org Lett 14(20):5334–5337 (j) Hartmann M, Li Y, Studer A (2012) Transition-metal-free oxyarylation of alkenes with aryl diazonium salts and tempona. J Am Chem Soc 134(40):16516–16519 (k) Ding YY, Cheng K, Qi CZ et al (2012) Ferrous salt-promoted homocoupling of arenediazonium tetrafluoroborates under mild conditions. Tetrahedron Lett 53(46):6269–6272
14. (a) Curran DP, Solovyev A, Brahmi MM et al (2011) Synthesis and reactions of n-heterocyclic carbene boranes. Angew Chem Int Ed 50(44):10294–10317 (b) Martin CD, Soleilhavoup M, Bertrand G (2013) Carbene-stabilized main group radicals and radical ions. Chem Sci 4 (8):3020–3030 (c) Lu DM, Wu C, Li PF (2014) Synergistic effects of lewis bases and substituents on the electronic structure and reactivity of boryl radicals. Chem-Eur J 20 (6):1630–1637 (d) Ueng SH, Fensterbank L, Lacote E et al (2010) Radical deoxygenation of xanthates and related functional groups with new minimalist n-heterocyclic carbene boranes. Org Lett 12(13):3002–3005 (e) Pan XC, Lacote E, Lalevee J et al (2012) Polarity reversal catalysis in radical reductions of halides by n-heterocyclic carbene boranes. J Am Chem Soc 134(12):5669–5674 (f) Pan XC, Vallet AL, Schweizer S et al (2013) Mechanistic and preparative studies of radical chain homolytic substitution reactions of n-heterocyclic carbene boranes and disulfides. J Am Chem Soc 135(28):1048410491
15. Zakharkin LI, Kovredov AI, Olshevskaya VA et al (1984) Synthesis of sigma-(ortho-carboran-9-yl)- and sigma-(meta-carboran-9-yl)-pi-cyclopentadienyldicarbonyliron and their rearrangement in reactions with bromine to pi-(ortho-carboran-9-yl)cyclopentadienyl and pi-(meta-carboran-9-yl)cyclopentadienyl-dicarbonyliron bromides, respectively. J Organomet Chem 267(1):81–91
16. Zharov I, Havlas Z, Orendt AM et al (2006) $CB_{11}Me_{11}$ boronium ylides: Carba-closo-dodecaboranes with a naked boron vertex. J Am Chem Soc 128(18):6089–6100
17. Ringstrand B, Kaszynski P (2013) Functionalization of the $[closo\text{-}1\text{-}CB_9H_{10}]^-$ anion for the construction of new classes of liquid crystals. Acc Chem Res 46(2):214–225

Chapter 8
Conclusion

In this thesis, we described our exploration on the functionalization of carboranes via *o*-carboryne intermediates. As shown in Scheme 8.1, our research initiated from the cage carbon functionalization of *o*-carborane via 1,2-dehydro-*o*-carborane (*o*-carboryne) intermediate, which could be produced by known methods. On the other hand, highly selective and efficient cage carbon functionalization of carborane was also realized via 1,3-dehydro-*o*-carborane, a reactive intermediate successfully generated from 3-diazonium-*o*-carborane tetrafluoroborate. With this precursor in hand, a simple and general cage boron arylation process has also been developed via metal-free photoredox catalysis.

First, we described a regioselective insertion of *o*-carborynes into α-C–H bond of amines. Different from our previous work on the reaction of *o*-carborynes with aliphatic ethers, the bonding form of *o*-carboryne is involved in this reaction. It undergoes the nucleophilic addition with tertiary amines to generate a zwitterionic intermediate, and then, an intramolecular proton transfer, followed by a nucleophilic attack, leads to the formation of the final α-C–H bond insertion product. This serves as a new methodology for the generation of a series of 1-aminoalkyl-*o*-carboranes.

On the other hand, *o*-carboryne is a very useful synthon for the synthesis of a variety of carborane-functionalized heterocycles. Reaction of *o*-carboryne with *N*-protected indoles gave carborane-fused indolines if the protecting group was TMS via dearomative [2 + 2] cycloaddition, or carboranyl indoles for *N*-alkyl ones through formal C–H insertion reaction. For *N*-aryl indoles, both reactions were observed, giving two products, in which the product ratio was dependent upon the nature of substituents on aryl rings. In general, electron-withdrawing substituents favor [2 + 2] cycloaddition, whereas electron-donating substituents promote C–H insertion pathway. A stepwise reaction mechanism was proposed to account for the experimental observations. These protocols offer general and efficient methods for the preparation of carborane-functionalized indoles and indolines as well as other heterocycles. The observed dearomative [2 + 2] cycloaddition also represents the

D. Zhao, *Functionalization of Carborane via Carboryne Intermediates*,
Springer Theses, DOI 10.1007/978-981-10-1569-4_8

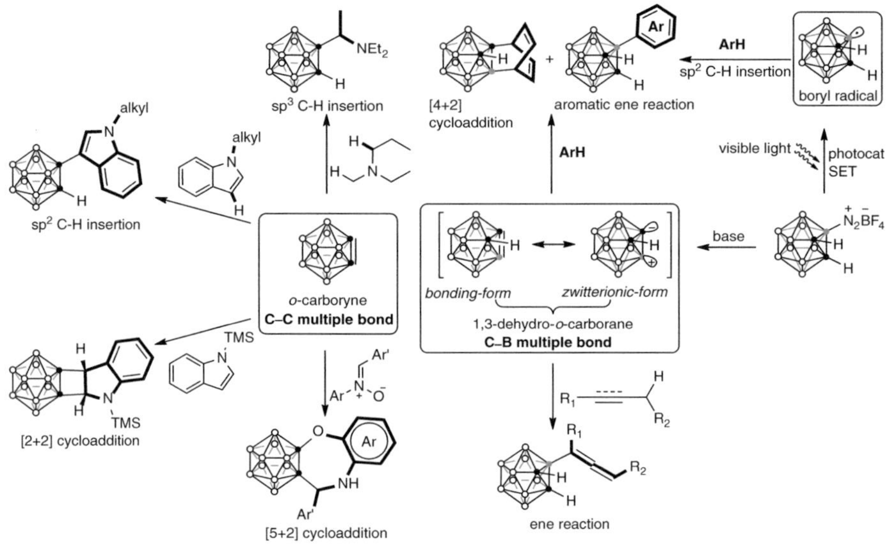

Scheme 8.1 Functionalization of *o*-carborane

first example for indoles to undergo such reaction in the absence of transition metals or without UV irradiation.

The following chapter describes the first example of transition-metal-free [5 + 2] cycloaddition of nitrone with *o*-carboryne. In this process, the nitrone moieties served as five-atom coupling partners with both the two heteroatoms incorporated into the frameworks of the final products. Mechanistic studies and DFT calculation suggest that the N–O bond cleavage in the initially formed [3 + 2] cycloadducts, followed by an intramolecular nucleophilic addition and rearomatization, leads to the final products. This approach allows rapid construction of carborane-fused seven-membered heterocycles with broad substrate scope, which also provides new insight into the chemistry of nitrones and *o*-carboryne.

Furthermore, we developed an efficient precursor for 1,3-dehydro-*o*-carborane that features a cage C–B multiple bond and such reactive intermediate serves as a very useful synthon for simultaneous functionalization of both cage carbon and boron vertices.

1,3-Dehydro-*o*-carborane undergoes Diels–Alder reaction with various arenes to give [4 + 2] cycloadducts, in which electronic factor governs the regioselectivity. Meanwhile, for arenes bearing benzylic C–H bonds, highly regioselective aromatic ene reaction was also observed with the product ratio of [4 + 2] cycloaddition over ene reaction depending on the number of benzylic protons. These results suggest that 1,3-dehydro-*o*-carborane is best described as a resonance hybrid of a bonding form and a zwitterionic form. Thus, it shares some chemical properties with those of *o*-carboryne; on the other hand, it also has its own unique properties due mainly to the polarized cage C–B bond.

In addition to the above aromatic ene reaction, 1,3-dehydro-*o*-carborane can readily react with a series of alkenes and alkynes in ene reaction fashions, providing a general, high-yielding synthesis of B(3)-alkenyl/allenyl carboranes with excellent regioselectivity. Ene reactions are dominant in these processes, which can be ascribed to the highly polarized cage C–B multiple bond of 1,3-dehydro-*o*-carborane. The experimental results reveal that the hydrogen shift in the ene reaction is a proton transfer in nature. The present work also demonstrates that 1,3-dehydro-*o*-carborane is a useful synthon for selective cage B-functionalization of *o*-carboranes. The newly prepared carboranyl alkenes and allenes may serve as useful starting materials for further functionalization.

3-Diazonium-*o*-carborane tetrafluoroborate serves not only as an efficient precursor for 1,3-dehydro-*o*-carborane, but also as an ideal source of carboranyl boryl radical. For instance, it can undergo regioselective C–H insertion with simple arenes with a wide range of functional groups. This general and simple procedure provides a transition-metal-free alternative for the synthesis of cage boron-arylated carboranes, which may found application in material science. This work also represents a synthetic useful application of boryl radicals.

The above works described in this thesis demonstrate that like the importance of benzyne in modern arene chemistry, *o*-carboryne intermediates are very useful synthons for functional *o*-carboranes. On the other hand, these intermediates have the unique properties of their own due to their steric bulkiness and high electron deficiency, which is different from that of benzyne.

Experimental Section

General Procedures All experiments were carried out in flame-dried glassware under an atmosphere of dry dinitrogen or argon using standard Schlenk techniques unless otherwise specified. All organic solvents were freshly distilled from Na-K alloy immediately prior to use. All other chemicals were purchased from either Aldrich or Acros Chemical Co. and used as received unless otherwise noted. 3-Phenyl-*o*-carborane, 3-chloro-*o*-carborane, 9,12-dimethyl-*o*-carborane, 9,12-diiodo-*o*-carborane, 4,5,7,8,9,10,11,12-octamethyl-*o*-carborane, *N*,*N*-diethylbenzylamine, *N*-benzylpyrrolidine, trimethylsilyldiethylamine, 1-Me-1,2-$C_2B_{10}H_{11}$, 1-TMS-1*H*-indoles, 1-methyl-1*H*-indoles, 1-allyl-1*H*-indole, 1-benzyl-1*H*-indole, 1-methyl-1*H*-benzimidazole, 1-aryl-1*H*-indole, 1,2-diboromo-*o*-carborane, and 1-*t*-butyldimethylsilyl-1*H*-indole, *C*-aryl-*N*-aryl-nitrones, 3-NH_2-*o*-carborane, 1,3-Diynes, 1,1,1-trifluoroundec-3-yne, and 1-Me_3Si-2-[IPh(OAc)]-1,2-$C_2B_{10}H_{10}$ were prepared according to literature methods. ^{1}H NMR spectra were recorded on a Bruker DPX 400 spectrometer at 400 MHz. ^{13}C NMR spectra were recorded on a Bruker DPX 400 spectrometer at 100 MHz. ^{11}B NMR spectra were recorded on a Bruker DPX 300 spectrometer at 96 MHz or a Varian Inova 400 spectrometer at 128 MHz. Chemical shifts were reported in ppm with reference to the residual solvent resonances of the deuterated solvents for proton and carbon chemical shifts and to external $BF_3{\cdot}OEt_2$ (0.0 ppm) for boron chemical shifts. The data were reported as follows: chemical shift, multiplicity (s = singlet, d = doublet, t = triplet, q = quadruplet, m = multiplet or unresolved, br = broad), coupling constant (s) in Hz, integration, and assignment. Mass spectra were obtained on a Thermo Finnigan MAT 95 XL spectrometer or Waters Micromass GCT Premier. Elemental analyses were performed by the Shanghai Institute of Organic Chemistry, Chinese Academic Society, China.

General Procedure for α-Carboranylation of Amines To a suspension of 1,2-Li_2-*o*-$C_2B_{10}H_{10}$ (1.0 mmol) in *n*-hexane (10 mL), prepared in situ from the reaction of ^{n}BuLi (1.25 mL, 1.6 M in *n*-hexane, 2.0 mmol) with *o*-carborane (144.2 mg, 1.0 mmol) at 0 °C, was added iodine (253.8 mg, 1.0 mmol) at room temperature. After stirring for 15 min at room temperature, iodine was completely disappeared and a white suspension was obtained. To this suspension was added 10 equiv (2.5 equiv for benzyl amines) of amine **II-4b-4t** via a syringe at room

D. Zhao, *Functionalization of Carborane via Carboryne Intermediates*,
Springer Theses, DOI 10.1007/978-981-10-1569-4

temperature, and the resulting mixture was stirred at room temperature for 15 min, heated at 60 °C for 5 h (1 h for **II-4b-4e**) and then quenched by wet *n*-hexane. After removal of solvents in vacuo, the residue was examined by ^{1}H NMR spectroscopy and then subjected to flash column chromatography on silica gel (230–400 mesh) using *n*-hexane as eluent to give desired product **II-5**.

1-(1-Diethylaminoethyl)-*o*-carborane **II-5aa**: Colorless oil. Yield: 72 %. ^{1}H NMR (400 MHz, $CDCl_3$): δ 4.08 (br, 1H) (cage C*H*), 3.43 (q, *J* = 6.8 Hz, 1H) (NC*H*), 2.52 (br, 2H), 2.31 (br, 2H) (NCH*H*), 1.23 (d, *J* = 6.8 Hz, 3H), 0.98 (br, 6H) (CH_3). ^{13}C{^{1}H} NMR (100 MHz, $CDCl_3$): δ 81.0 (cage *C*), 60.0, 59.3 (cage *C*), 46.8, 41.3, 14.8, 19.8, 12.9. ^{11}B{^{1}H} NMR (128 MHz, $CDCl_3$): δ −3.6 (1B), −5.2 (1B), −9.6 (2B), −11.1 (1B), −12.0 (1B), −12.6 (1B), −13.3 (1B), −14.3 (2B). HRMS (EI) calcd for $C_8H_{25}^{11}B_8^{10}B_2N^+$: 243.3103. Found 243.3101.

1-(1-Dipropylaminopropyl)-*o*-carborane **II-5b**: Colorless crystals. Yield: 65 %. ^{1}H NMR (400 MHz, $CDCl_3$): δ 4.18 (br, 1H) (cage C*H*), 3.22 (dd, *J* = 8.4, 6.0 Hz, 1H) (NC*H*), 2.63 (m, 1H), 2.52 (m, 2H), 2.40 (m, 1H) (NCH_2), 1.75 (m, 2H), 1.49 (m, 2H), 1.32 (m, 2H) (CH_2), 1.00 (t, *J* = 7.6 Hz, 3H), 0.83 (m, 6H) (CH_3). ^{13}C{^{1}H} NMR (100 MHz, $CDCl_3$): δ 81.3 (cage *C*), 68.0, 59.8 (cage *C*), 58.6, 50.2, 25.6, 22.8, 22.7, 13.4, 11.8, 11.6. ^{11}B{^{1}H} NMR (128 MHz, $CDCl_3$): δ −4.0 (1B), −5.2 (1B), −9.4 (1B), −10.0 (1B), −12.2 (3B), −13.5 (1B), −14.7 (2B). HRMS (EI) calcd for $C_{11}H_{31}^{11}B_8^{10}B_2N^+$: 285.3460. Found 285.3462.

1-(1-Dibutylaminobutyl)-*o*-carborane **II-5c**: Colorless oil. Yield: 60 %. ^{1}H NMR (400 MHz, $CDCl_3$): δ 4.19 (br, 1H) (cage C*H*), 3.28 (dd, *J* = 8.0, 5.6 Hz, 1H) (NC*H*), 2.54 (m, 3H), 2.35 (m, 1H), (NCH_2), 1.65 (m, 2H), 1.41 (m, 4H), 1.26 (m, 6H) (CH_2), 0.93 (m, 9H) (CH_3). ^{13}C{^{1}H} NMR (100 MHz, $CDCl_3$): δ 81.3 (cage *C*), 65.8, 60.0 (cage *C*), 55.7, 48.4, 34.3, 31.8, 31.5, 21.8, 20.7, 20.5, 14.2, 14.2, 14.1. ^{11}B{^{1}H} NMR (128 MHz, $CDCl_3$): δ −4.1 (1B), −5.2 (1B), −9.4 (1B), −9.8 (1B), −12.0 (3B), −13.6 (1B), −14.6 (2B). HRMS (EI) calcd for $C_{14}H_{37}^{11}B_8^{10}B_2N^+$: 327.3931. Found 327.3932.

1-[1-Bis(2-methylpropyl)amino-2-methylpropyl]-*o*-carborane **II-5d**: Colorless crystals. Yield: 61 %. ^{1}H NMR (400 MHz, $CDCl_3$): δ 4.24 (br, 1H) (cage C*H*), 3.39 (d, *J* = 4.0 Hz, 1H) (NC*H*), 2.65 (m, 1H), 2.56 (m, 2H), 2.44 (m, 1H) (NCH_2), 2.20 (m, 1H), 1.70 (m, 2H) (C*H*), 1.17 (dd, *J* = 6.8, 4.4 Hz, 6H), 0.91 (m, 12H) (CH_3). ^{13}C{^{1}H} NMR (100 MHz, $CDCl_3$): δ 80.0 (cage *C*), 70.4, 63.9, 60.8 (cage *C*), 59.8, 33.2, 28.3, 27.6, 23.4, 21.7, 21.3, 21.3, 21.2, 20.6. ^{11}B{^{1}H} NMR (128 MHz, $CDCl_3$): δ −4.3 (2B), −9.7 (2B), −11.1 (2B), −14.2 (4B). HRMS (EI) calcd for $C_{14}H_{37}^{11}B_8^{10}B_2N^+$: 327.3931. Found 327.3936.

1-[(1-Bis(2-propenylamino)-2-propenyl)-*o*-carborane **II-5e**: Colorless oil. Yield: 45 %. ^{1}H NMR (400 MHz, C_6D_6): δ 5.45 (m, 2H), 5.23 (m, 1H), 4.92 (m, 5H), 4.65 (dd, *J* = 16.8, 0.8 Hz, 1H) (olefinic C*H*), 3.52 (d, *J* = 10.0 Hz, 1H) (NC*H*), 3.47 (br, 1H) (cage C*H*), 3.04 (m, 2H), 2.20 (dd, *J* = 14.0, 8.4 Hz, 2H) (NCH_2). ^{13}C{^{1}H} NMR (100 MHz, C_6D_6): δ 135.6, 130.8, 121.8, 118.5, 78.5 (cage *C*), 65.8, 59.1, 54.0 (cage *C*). ^{11}B{^{1}H} NMR (128 MHz, C_6D_6): δ −2.5 (1B), −4.1

(1B), −9.2 (2B), −12.1 (3B), −13.4 (3B). HRMS (EI) calcd for $C_{11}H_{25}^{11}B_{8}^{10}B_{2}N^{+}$: 279.2990. Found 279.2995.

1-[1-Bis(2-methylpropyl)aminoethyl]-*o*-carborane **II-5f**: Colorless crystals. Yield: 32 %. ^{1}H NMR (400 MHz, $CDCl_3$): δ 3.95 (br, 1H) (cage C*H*), 3.62 (q, J = 6.8 Hz, 1H) (NC*H*), 3.15 (m, 2H) (NC*H*), 1.38 (d, J = 7.2 Hz, 3H), 1.01 (m, 12H) (CH_3). $^{13}C\{^{1}H\}$ NMR (100 MHz, $CDCl_3$): δ 82.1 (cage *C*), 58.9 (cage *C*), 56.3, 47.3, 45.0, 25.3, 23.1, 21.4, 20.9, 20.0. $^{11}B\{^{1}H\}$ NMR (128 MHz, $CDCl_3$): δ −7.5 (1B), −9.4 (1B), −13.9 (1B), −14.6 (2B), −16.4 (2B), −17.2 (1B), −18.6 (2B). HRMS (EI) calcd for $C_{10}H_{29}^{11}B_{8}^{10}B_{2}N^{+}$: 271.2867. Found 271.2864.

1-(Methylbutylaminomethyl)-*o*-carborane **II-5ga**: Colorless oil. Yield: 77 %. ^{1}H NMR (400 MHz, $CDCl_3$): δ 3.97 (br, 1H) (cage C*H*), 3.06 (s, 2H) (NCH_2), 2.44 (t, J = 7.2 Hz, 2H) (NCH_2), 2.31 (s, 3H) (NCH_3), 1.38 (m, 2H), 1.29 (m, 2H) (CH_2), 0.91 (t, J = 7.2 Hz, 3H) (CH_3). $^{13}C\{^{1}H\}$ NMR (100 MHz, $CDCl_3$): δ 75.9 (cage *C*), 62.6, 59.4 (cage *C*), 58.1, 43.9, 29.9, 20.4, 14.1. $^{11}B\{^{1}H\}$ NMR (128 MHz, $CDCl_3$): δ −3.5 (1B), −6.0 (1B), −9.6 (2B), −11.9 (2B), −13.1 (2B), −13.8 (2B). HRMS (EI) calcd for $C_{8}H_{25}^{11}B_{8}^{10}B_{2}N^{+}$: 243.3103. Found 243.3109.

1-(Dibutylaminomethyl)-*o*-carborane **II-5ha**: Colorless oil. Yield: 41 %. ^{1}H NMR (400 MHz, $CDCl_3$): δ 3.89 (br, 1H) (cage C*H*), 3.10 (s, 2H) (NCH_2), 2.46 (m, 4H) (NCH_2), 1.34 (m, 4H), 1.26 (m, 4H) (CH_2), 0.91 (t, J = 7.2 Hz, 6H) (CH_3). $^{13}C\{^{1}H\}$ NMR (100 MHz, $CDCl_3$): δ 60.4, 57.8 (cage *C*), 55.0, 29.4, 20.6, 14.2, another cage *C* was overlapped with the solvent. $^{11}B\{^{1}H\}$ NMR (128 MHz, $CDCl_3$): δ −3.1 (1B), −5.6 (2B), −9.3 (2B), −11.5 (2B), −12.6 (1B), −13.4 (2B). HRMS (EI) calcd for $C_{11}H_{31}^{11}B_{8}^{10}B_{2}N^{+}$: 285.3460. Found 285.3461.

1-(Methylbutylaminobutyl)-*o*-carborane **II-5hb**: Data were collect from the mixture of **II-5ha** and **II-5hb**. Yield: 22 %. ^{1}H NMR (400 MHz, $CDCl_3$): δ 4.09 (br, 1H) (cage C*H*), 3.25 (q, J = 4.8 Hz, 1H) (NC*H*), 2.65 (m, 2H) (NCH_2), 2.58 (m, 2H) (CH_2), 2.23 (s, 3H) (NCH_3), 1.65 (m, 4H) (CH_2), 0.93 (m, 6H) (CH_3). $^{13}C\{^{1}H\}$ NMR (100 MHz, $CDCl_3$): δ 81.5 (cage *C*), 68.5 (cage *C*), 67.5, 60.6, 60.0, 34.2, 33.7, 21.6, 20.3, 14.1.

1-(Diethylaminobenzyl)-*o*-carborane **II-5i**: Colorless oil. Yield: 75 %. ^{1}H NMR (400 MHz, $CDCl_3$): δ 7.36 (m, 3H), 7.24 (m, 2H) (aromatic C*H*), 4.42 (s, 1H) (NC*H*), 3.86 (br, 1H) (cage C*H*), 2.78 (m, 2H), 2.15 (m, 2H) (NCH_2), 1.03 (m, 6H) (CH_3). $^{13}C\{^{1}H\}$ NMR (100 MHz, $CDCl_3$): δ 133.2, 130.7, 128.8, 128.4, 78.9 (cage *C*), 70.2, 59.7 (cage *C*), 44.7, 13.7. $^{11}B\{^{1}H\}$ NMR (128 MHz, $CDCl_3$): δ −3.3 (1B), −4.6 (1B), −9.5 (2B), −11.3 (3B), −13.7 (3B). HRMS (EI) calcd for $C_{13}H_{25}^{11}B_{8}^{10}B_{2}N^{+}$ $([M-2H]^{+})$: 304.3148. Found 304.3067.

1-(1-Benzylpyrrolidin-1-yl)-*o*-carborane **II-5j**: Colorless oil. Yield: 58 %. ^{1}H NMR (400 MHz, $CDCl_3$): δ 7.37 (m, 3H), 7.24 (m, 2H) (aromatic C*H*), 4.48 (s, 1H) (NC*H*), 3.96 (br, 1H) (cage C*H*), 2.60 (m, 2H), 2.52 (m, 2H) (NCH_2), 1.64 (m, 4H) (CH_2). $^{13}C\{^{1}H\}$ NMR (100 MHz, $CDCl_3$): δ 130.0, 130.7, 128.9, 128.4, 79.4 (cage *C*), 71.2, 60.2 (cage *C*), 51.5, 23.2. $^{11}B\{^{1}H\}$ NMR (128 MHz, $CDCl_3$): δ −3.8 (1B), −4.5 (1B), −9.1 (2B), −11.3 (3B), −13.6 (3B). HRMS (EI) calcd for $C_{13}H_{25}^{11}B_{8}^{10}B_{2}N^{+}$: 303.2991. Found 303.2990.

1-(1-Diethylamino-4-fluorobenzyl)-*o*-carborane **II-5k**: Colorless oil. Yield: 83 %. ^{1}H NMR (400 MHz, CDCl$_3$): δ 7.22 (m, 2H), 7.06 (m, 2H) (aromatic C*H*), 4.42 (s, 1H) (NC*H*), 3.89 (br, 1H) (cage C*H*), 2.75 (m, 2H), 2.12 (m, 2H) (NCH_2), 1.03 (t, J = 6.8 Hz, 6H) (CH_3). ^{13}C{^{1}H} NMR (100 MHz, CDCl$_3$): δ 162.9 ($J_{C\text{–}F}$ = 252.0 Hz), 132.3 ($J_{C\text{–}F}$ = 8.0 Hz), 129.1 ($J_{C\text{–}F}$ = 4.0 Hz), 115.4 ($J_{C\text{–}F}$ = 21.0 Hz), 78.7 (cage *C*), 69.3, 59.7 (cage *C*), 44.6, 13.7. ^{11}B{^{1}H} NMR (128 MHz, CDCl$_3$): δ −3.3 (1B), −4.6 (1B), −9.5 (2B), −11.3 (3B), −13.6 (3B). HRMS (EI) calcd for $C_{13}H_{26}F^{11}B_8^{10}B_2N^+$: 323.2896. Found 323.2890.

1-(1-Diethylamino-4-methoxylbenzyl)-*o*-carborane **II-5l**: Colorless oil. Yield: 67 %. ^{1}H NMR (400 MHz, CDCl$_3$): δ 7.15 (d, J = 8.8 Hz, 2H), 6.87 (d, J = 8.8 Hz, 2H) (aromatic C*H*), 4.38 (s, 1H) (NC*H*), 3.85 (br, 1H) (cage C*H*), 3.81 (s, 3H) (OCH_3), 2.75 (m, 2H), 2.13 (m, 2H) (NCH_2), 1.02 (m, 6H) (CH_3). ^{13}C{^{1}H} NMR (100 MHz, CDCl$_3$): δ 159.8, 131.9, 125.4, 113.7, 79.3 (cage *C*), 69.5, 59.7 (cage *C*), 55.4, 44.6, 13.7. ^{11}B{^{1}H} NMR (128 MHz, CDCl$_3$): δ −3.2 (1B), −4.7 (1B), −9.5 (2B), −11.2 (3B), −13.6 (3B). HRMS (EI) calcd for $C_{14}H_{29}^{11}B_8^{10}B_2N^+$: 320.3573. Found 320.3573.

1-(1-Ethylpyrrolidin-2-yl)-*o*-carborane **II-5ma**: Colorless oil. Yield: 69 %. ^{1}H NMR (400 MHz, CDCl$_3$): δ 4.03 (br, 1H) (cage C*H*), 3.35 (dd, J = 8.0, 2.8 Hz, 1H) (NC*H*), 3.07 (m, 1H), 2.73 (m, 1H), 2.55 (m, 1H), 2.40 (m, 1H) (NCH*H*), 1.98 (m, 2H), 1.76 (m, 2H) (CH_2), 1.04 (t, J = 7.2 Hz, 3H) (CH_3). ^{13}C{^{1}H} NMR (100 MHz, CDCl$_3$): δ 81.8 (cage *C*), 68.2, 59.3 (cage *C*), 54.1, 52.2, 33.2, 25.2, 14.1. ^{11}B{^{1}H} NMR (128 MHz, CDCl$_3$): δ −3.7 (1B), −5.2 (1B), −9.2 (1B), −9.7 (1B), −12.0 (3B), −13.3 (1B), −14.2 (2B). HRMS (EI) calcd for $C_8H_{23}^{11}B_8^{10}B_2N^+$: 241.2946. Found 241.2943.

1-(1-Ethylpyrrolidin-1-yl)-*o*-carborane **II-5mb**: Data were collect from the mixture of **II-5ma** and **II-5mb**. Yield: 9 %. ^{1}H NMR (400 MHz, CDCl$_3$): δ 4.20 (br, 1H) (cage C*H*), 3.56 (m, 1H) (NC*H*), 3.37 (m, 1H), 2.73 (m, 1H), 2.55 (m, 1H), 2.40 (m, 1H) (NCH*H*), 1.98 (m, 2H), 1.76 (m, 2H) (CH_2), 1.21 (d, J = 6.8 Hz, 3H) (CH_3). ^{13}C{^{1}H} NMR (100 MHz, CDCl$_3$): δ 81.8 (cage *C*), 68.2, 60.0 (cage *C*), 58.9, 25.2, 14.3.

1-(1-Methylpiperidin-1-yl)-*o*-carborane **II-5n**: Colorless oil. Yield: 65 %. ^{1}H NMR (400 MHz, CDCl$_3$): δ 3.99 (br, 1H) (cage C*H*), 2.98 (s, 2H) (NCH_2), 1.52 (br, 5H) (CH_2), other C*H* were overlapped with carborane BH. ^{13}C{^{1}H} NMR (100 MHz, CDCl$_3$): δ 75.6 (cage *C*), 62.9, 58.2 (cage *C*), 56.3, 26.3, 23.7. ^{11}B{^{1}H} NMR (128 MHz, CDCl$_3$): δ −3.4 (1B), −5.8 (1B), −9.3 (2B), −11.7 (2B), −12.8 (2B), −13.6 (2B). HRMS (EI) calcd for $C_8H_{23}^{11}B_8^{10}B_2N^+$: 241.2946. Found 241.2948.

1-(1-Piperidinylethyl)-*o*-carborane **II-5o**: Colorless oil. Yield: 65 %. ^{1}H NMR (400 MHz, CDCl$_3$): δ 4.15 (br, 1H) (cage C*H*), 3.24 (q, J = 7.2 Hz, 1H) (NC*H*), 1.21 (d, J = 7.2 Hz, 3H) (CH_3), other C*H* were overlapped with carborane BH. ^{13}C{^{1}H} NMR (100 MHz, CDCl$_3$): δ 80.9 (cage *C*), 63.9, 59.6 (cage *C*), 56.6, 47.5, 26.8, 26.2, 24.3, 13.7. ^{11}B{^{1}H} NMR (128 MHz, CDCl$_3$): δ −3.9 (1B), −5.3 (1B), −9.5 (2B), −11.6 (2B), −13.0 (2B), −14.2 (2B). HRMS (EI) calcd for $C_9H_{25}^{11}B_8^{10}B_2N^+$: 255.2989. Found 255.2987.

1-(1-Morpholinylethyl)-*o*-carborane **II-5p**: Colorless oil. Yield: 71 %. ^{1}H NMR (400 MHz, $CDCl_3$): δ 4.13 (br, 1H) (cage C*H*), 3.69 (br, 4H) (OCH_2), 3.23 (q, J = 7.2 Hz, 1H) (NC*H*), 2.64 (br, 2H), 2.43 (m, 2H) (NCH_2), 1.25 (d, J = 7.2 Hz, 3H) (CH_3). ^{13}C{^{1}H} NMR (100 MHz, $CDCl_3$): δ 80.1 (cage *C*), 67.2, 63.6 (cage *C*), 59.4, 13.9. ^{11}B{^{1}H} NMR (128 MHz, $CDCl_3$): δ −3.6 (1B), −5.0 (1B), −9.3 (2B), −11.6 (2B), −12.7 (2B), −14.0 (2B). HRMS (EI) calcd for $C_8H_{23}^{11}B_8^{10}B_2NO^+$: 271.2983. Found 271.2987.

1-(1-Ethylaminoethyl)-*o*-carborane **II-5q**: Colorless oil. Yield: 68 %. ^{1}H NMR (400 MHz, $CDCl_3$): δ 4.44 (br, 1H) (cage C*H*), 3.22 (q, J = 6.4 Hz, 1H) (NC*H*), 2.78 (m, 1H), 2.52 (m, 1H) (NCH*H*), 1.30 (d, J = 6.4 Hz, 3H), 1.05 (t, J = 7.2 Hz, 3H) (CH_3). ^{13}C{^{1}H} NMR (100 MHz, $CDCl_3$): δ 80.7 (cage *C*), 59.8, 56.7 (cage *C*), 43.4, 21.8, 15.7. ^{11}B{^{1}H} NMR (128 MHz, $CDCl_3$): δ −3.7 (1B), −5.1 (1B), −8.8 (1B), −9.8 (1B), −11.2 (1B), −12.3 (2B), −13.0 (1B), −14.6 (1B), −15.2 (1B). HRMS (EI) calcd for $C_6H_{21}^{11}B_8^{10}B_2N^+$: 215.2389. Found 215.2383.

1-(Bistrimethylsilylaminomethyl)-*o*-carborane **II-5r**: Colorless oil. Yield: 73 %. ^{1}H NMR (400 MHz, C_6D_6): δ 3.16 (br, 1H) (cage C*H*), 3.16 (s, 2H) (NCH_2), −0.01 (s, 18H) (CH_3). ^{13}C{^{1}H} NMR (100 MHz, C_6D_6): δ 79.5 (cage *C*), 59.7, 52.5 (cage *C*), 2.7. ^{11}B{^{1}H} NMR (128 MHz, $CDCl_3$): δ −2.8 (1B), −5.6 (1B), −10.5 (2B), −12.3 (2B), −14.0 (4B). HRMS (EI) calcd for $C_9H_{31}^{11}B_8^{10}B_2NSi_2^+$: 319.4320. Found 319.4318.

General Procedure for α-Carboranylation of Triethylamine. A Representative Procedure To a suspension of dilithiocarborane (1.0 mmol) in *n*-hexane (10 mL), prepared in situ from the reaction of ^{n}BuLi (1.25 mL, 1.6 M in *n*-hexane, 2.0 mmol) with *o*-carboranes (**II-1a-1f**) (1.0 mmol) at 0 °C, was added iodine (253.8 mg, 1.0 mmol) at room temperature. After stirring for 15 min at room temperature, iodine was completely disappeared and a white suspension was obtained. To this suspension was added 10 equiv of triethylamine **II-4a** via a syringe at room temperature, and the resulting mixture was stirred at room temperature for 15 min, heated at 60 °C for 5 h and then quenched by wet *n*-hexane. After removal of solvents in vacuo, the residue was examined by ^{1}H NMR spectroscopy and then subjected to flash column chromatography on silica gel (230–400 mesh) using *n*-hexane as eluent to give desired product **II-5**.

1-(1-Diethylaminoethyl)-3-phenyl-*o*-carborane **II-5ba**: Colorless oil. Yield: 39 %. The relative configuration was not determined. ^{1}H NMR (400 MHz, $CDCl_3$): δ 7.44 (m, 2H), 7.40 (m, 3H) (aromatic C*H*), 4.07 (br, 1H) (cage C*H*), 2.98 (q, J = 7.2 Hz, 1H) (NC*H*), 2.53 (m, 1H), 2.36 (m, 1H), 1.99 (m, 1H), 1.86 (m, 1H) (NCH*H*), 1.12 (d, J = 6.8 Hz, 3H), 1.09 (m, 3H), 0.51 (m, 3H) (CH_3). ^{13}C{^{1}H} NMR (100 MHz, $CDCl_3$): δ 134.1, 129.8, 128.5, 81.5 (cage *C*), 58.9, 57.6 (cage *C*), 45.9, 40.7, 14.8, 13.6, 12.0. ^{11}B{^{1}H} NMR (128 MHz, $CDCl_3$): δ −2.0 (1B, *B*Ph), −4.0 (2B), −8.9 (2B), −12.8 (3B), −14.1 (2B). HRMS (EI) calcd for $C_{14}H_{29}^{11}B_8^{10}B_2N^+$: 319.3305. Found 319.3304.

1-(1-Diethylaminoethyl)-3-phenyl-*o*-carborane **II-5ba′**: Colorless oil. Yield: 39 %. The relative configuration was not determined. ^{1}H NMR (400 MHz, C_6D_6):

δ 7.14 (m, 2H), 7.09 (m, 3H) (aromatic C*H*), 3.15 (br, 1H) (cage C*H*), 2.93 (q, *J* = 6.8 Hz, 1H) (NC*H*), 2.20 (br, 2H), 1.83 (m, 2H), (NCH*H*), 0.76 (br, 3H), 0.62 (br, 3H), 0.37 (d, *J* = 6.8 Hz, 3H) (CH_3). $^{13}C\{^1H\}$ NMR (100 MHz, C_6D_6): δ 134.1, 129.8, 128.5, 81.3 (cage *C*), 57.7, 57.2 (cage *C*), 45.8, 41.1, 13.4, 13.0, 12.9. $^{11}B\{^1H\}$ NMR (128 MHz, C_6D_6): δ −3.8 (3B, *B*Ph was overlapped), −8.1 (1B), −10.8 (1B), −12.5 (2B), −14.1 (3B). HRMS (EI) calcd for $C_{14}H_{29}^{11}B_8^{10}B_2N^+$: 319.3305. Found 319.3302.

1-(1-Diethylaminoethyl)-3-chloro-*o*-carborane **II-5ca**: Colorless oil. Yield: 40 %. The relative configuration was not determined. ^1H NMR (400 MHz, $CDCl_3$): δ 3.91 (br, 1H) (cage C*H*), 3.72 (q, *J* = 6.8 Hz, 1H) (NC*H*), 2.69 (m, 1H), 2.61 (m, 2H), 2.41 (m, 1H) (NCH*H*), 1.30 (d, *J* = 7.2 Hz, 3H), 0.99 (br, 6H) (CH_3). $^{13}C\{^1H\}$ NMR (100 MHz, $CDCl_3$): δ 81.3 (cage *C*), 61.4, 57.7 (cage *C*), 46.2, 40.8, 15.4, 13.6, 12.2. $^{11}B\{^1H\}$ NMR (128 MHz, $CDCl_3$): δ −3.1 (1B, *B*Cl), −4.2 (1B), −4.9 (1B), −9.4 (1B), −12.9 (4B), −14.3 (1B), 15.1 (1B). HRMS (EI) calcd for $C_8H_{24}^{11}B_8^{10}B_2NCl^+$: 277.6794. Found 277.6793.

1-(1-Diethylaminoethyl)-3-chloro-*o*-carborane **II-5ca′**: Colorless oil. Yield: 40 %. The relative configuration was not determined. ^1H NMR (400 MHz, C_6D_6): δ 3.77 (m, 1H) (NC*H*), 3.58 (br, 1H) (cage C*H*), 2.70 (m, 2H), 2.50 (m, 2H), (NCH*H*), 1.18 (d, *J* = 6.8 Hz, 3H), 1.03 (br, 6H), (CH_3). $^{13}C\{^1H\}$ NMR (100 MHz, C_6D_6): δ 60.1 (cage *C*), 58.6, 46.4, 41.1, 13.8, 12.9, another cage *C* was not observed. $^{11}B\{^1H\}$ NMR (128 MHz, C_6D_6): δ −3.8 (3B, *B*Cl was overlapped), −8.1 (1B), −10.8 (1B), −12.5 (2B), −14.1 (3B). HRMS (EI) calcd for $C_8H_{24}^{11}B_8^{10}B_2NCl^+$: 277.6794. Found 277.6797.

1-(1-Diethylaminoethyl)-9,12-dimethyl-*o*-carborane **II-5da**: Colorless oil. Yield: 76 %. ^1H NMR (400 MHz, $CDCl_3$): δ 3.85 (br, 1H) (cage C*H*), 3.43 (m, 1H) (NC*H*), 2.52 (br, 4H) (NCH*H*), 1.22 (d, *J* = 7.2 Hz, 3H), 0.98 (m, 6H) (CH_3), 0.17 (m, 6H) (BCH_3). $^{13}C\{^1H\}$ NMR (100 MHz, $CDCl_3$): δ 73.8 (cage *C*), 59.4, 52.7 (cage *C*), 46.7, 41.5, 14.9, 13.9, 13.1. $^{11}B\{^1H\}$ NMR (128 MHz, $CDCl_3$): δ 6.2 (1B, *B*Me), 4.7 (1B, *B*Me), −7.7 (2B), −12.5 (3B), −13.4 (1B), −14.3 (1B), 15.7 (1B). HRMS (EI) calcd for $C_{10}H_{29}^{11}B_8^{10}B_2N^+$: 271.3303. Found 271.3307.

Control Experiments

Reaction of 1-I-2-Li-*o*-carborane with Triethylamine in the Dark To a suspension of 1,2-Li_2-*o*-$C_2B_{10}H_{10}$ (1.0 mmol) in *n*-hexane (10 mL) in the dark (covered with aluminum foil), prepared in situ from the reaction of ^{n}BuLi (1.25 mL, 1.6 M in *n*-hexane, 2.0 mmol) with *o*-carborane (144.2 mg, 1.0 mmol) at 0 °C, was added iodine (253.8 mg, 1.0 mmol) at room temperature. After stirring for 15 min at room temperature, iodine was completely disappeared and a white suspension was obtained. To this suspension was added 10 equiv of triethylamine **II-4a** via a syringe at room temperature, and the resulting mixture was stirred at room temperature for 15 min, heated at 60 °C for 1 h in the dark and then quenched by wet

n-hexane. After removal of solvents in vacuo, the residue was examined by ^{1}H NMR spectroscopy and then subjected to flash column chromatography on silica gel (230–400 mesh) using n-hexane as eluent to give desired product **II-5aa** (175.8 mg, 72 %).

Reaction of 1-I-2-Li-*o*-carborane with Triethylamine under UV light Irradiation (365 nm) To a suspension of 1,2-Li_2-*o*-$C_2B_{10}H_{10}$ (1.0 mmol) in n-hexane (10 mL), prepared in situ from the reaction of ^{n}BuLi (1.25 mL, 1.6 M in n-hexane, 2.0 mmol) with *o*-carborane (144.2 mg, 1.0 mmol) at 0 °C, was added iodine (253.8 mg, 1.0 mmol) at room temperature. After stirring for 15 min at room temperature, iodine was completely disappeared and a white suspension was obtained. To this suspension was added 10 equiv of triethylamine **II-4a** via a syringe at room temperature, and the resulting mixture was stirred at room temperature for 15 min, heated at 60 °C for 1 h under a handheld UV lamp irradiation (365 nm) and then quenched by wet n-hexane. After removal of solvents in vacuo, the residue was examined by ^{1}H NMR spectroscopy and then subjected to flash column chromatography on silica gel (230–400 mesh) using n-hexane as eluent to give desired product **II-5aa** (175.3 mg, 72 %).

Reaction of 1-I-2-Li-*o*-carborane with Triethylamine in the Presence of TEMPO To an in situ-prepared n-hexane suspension of 1-iodo-2-Li-*o*-$C_2B_{10}H_{10}$ (1.0 mmol) was added TEMPO (171.6 mg, 1.1 mmol) at 0 °C. Then, to this mixture was added 10 equiv of triethylamine **II-4a** via a syringe at room temperature, and the resulting mixture was stirred at room temperature for 15 min, heated at 60 °C for 5 h and then quenched by wet n-hexane. After removal of solvents in vacuo, the residue was examined by ^{1}H NMR spectroscopy and then subjected to flash column chromatography on silica gel (230–400 mesh) using n-hexane as eluent to give desired product **II-5aa** (150.7 mg, 62 %). The ^{1}H and GC-MS results showed that most of the crude product was **II-5aa**, with a small amount of the starting material 1-iodo-*o*-carborane coupled with a trace amount of carborane. No any TEMPO-coupled product was detected.

Reaction of 1-I-7-Li-*m*-carborane with Triethylamine under Standard Condition To an in situ-prepared n-hexane suspension of 1,7-Li_2-*o*-$C_2B_{10}H_{10}$ (0.5 mmol), prepared in situ from the reaction of ^{n}BuLi (0.625 mL, 1.6 M in n-hexane, 1.0 mmol) with *m*-carborane (77.1 mg, 0.5 mmol) at 0 °C, was added iodine (127.4 mg, 0.5 mmol) at room temperature. After stirring for 15 min at room temperature, iodine was completely disappeared and a white suspension was obtained. To this mixture was added 10 equiv of triethylamine **II-4a** via a syringe at room temperature, and the resulting mixture was stirred at room temperature for 15 min, heated at 60 °C for 5 h and then quenched by wet n-hexane. After removal of solvents in vacuo, the residue was examined by ^{1}H NMR spectroscopy and then subjected to flash column chromatography on silica gel (230–400 mesh) using n-hexane as eluent. The ^{1}H, ^{11}B NMR, and GC-MS results showed that most of the crude product was *m*-carborane.

Kinetic Isotope Effect Experiments

(1) **Reaction of 1-I-2-Li-*o*-carborane with Benzylpyrolidine-d_2.** To a suspension of 1,2-Li_2-*o*-$C_2B_{10}H_{10}$ (1.0 mmol) in *n*-hexane (10 mL), prepared in situ from the reaction of ^{n}BuLi (1.25 mL, 1.6 M in *n*-hexane, 2.0 mmol) with *o*-carborane (144.2 mg, 1.0 mmol) at 0 °C, was added iodine (253.8 mg, 1.0 mmol) at room temperature. After stirring for 15 min at room temperature, iodine was completely disappeared and a white suspension was obtained. To this suspension was added 2.5 equiv of benzylpyrolidine-d_2 **II-4j-D** (>99 %-D) via a syringe at room temperature, and the resulting mixture was stirred at room temperature for 15 min, heated at 60 °C for 1 h, and then quenched by wet *n*-hexane. After removal of solvents in vacuo, the residue was examined by ^{1}H NMR spectroscopy and then subjected to flash column chromatography on silica gel (230–400 mesh) using *n*-hexane as eluent to give desired product **II-5j-D$_2$** (160.7 mg, 53 %).

(2) **Reaction of 1-I-2-Li-*o*-carborane with Benzylpyrolidine/Benzylpyrolidine-d_2.** To a suspension of 1,2-Li_2-*o*-$C_2B_{10}H_{10}$ (1.0 mmol) in *n*-hexane (10 mL), prepared in situ from the reaction of ^{n}BuLi (1.25 mL, 1.6 M in *n*-hexane, 2.0 mmol) with *o*-carborane (144.2 mg, 1.0 mmol) at 0 °C, was added iodine (253.8 mg, 1.0 mmol) at room temperature. After stirring for 15 min at room temperature, iodine was completely disappeared and a white suspension was obtained. To this suspension was added 2.5 equiv of benzylpyrolidine/benzylpyrolidine-d_2 (>99 %-D) (1:1 in v:v) via a syringe at room temperature, and the resulting mixture was stirred at room temperature for 15 min, heated at 60 °C for 1 h and then quenched by wet *n*-hexane. After removal of solvents in vacuo, the residue was examined by ^{1}H NMR spectroscopy and then subjected to flash column chromatography on silica gel (230–400 mesh) using *n*-hexane as eluent to give desired product **II-5j** and its deuterated isomers (172.9 mg, 57 %).

Preparation of 1-Br-2-Li-o-$C_2B_{10}H_{10}$ in situ To a suspension of 1,2-Li_2-*o*-$C_2B_{10}H_{10}$ (0.5 mmol) in *n*-hexane (10 mL), prepared in situ from the reaction of ^{n}BuLi (0.625 mL, 1.6 M in *n*-hexane, 1.0 mmol) with *o*-carborane (72.1 mg, 0.5 mmol) at 0 °C, was added 1,2-Br_2-*o*-$C_2B_{10}H_{10}$ (0.5 mmol) (150.8 mg, 0.5 mmol) at room temperature. After stirring for 12 h at room temperature, a white suspension was obtained. ^{1}H, ^{11}B NMR, and GC-MS analysis of the hydrolyzed sample of the above suspension showed that most of the crude product was 1-Br-*o*-$C_2B_{10}H_{11}$ (>95 %).

General Procedure for [2 + 2] Cycloaddition of *N*-TMS Indoles with *o*-Carboryne. A Representative Procedure To an in situ-prepared suspension of 1-Br-2-Li-*o*-$C_2B_{10}H_{10}$ (1.0 mmol) in *n*-hexane (10 mL) was added 1.1 equiv of *N*-TMS indoles at room temperature. After stirring for 15 min at room temperature, the resulting mixture was heated at 60 °C for 6 h. The inorganic salt was removed by filtration and washed with *n*-hexane (5 mL × 3). After removal of solvents in vacuo, the residue was examined by ^{1}H NMR spectroscopy and then purified by flash column chromatography on silica gel (230–400 mesh) using *n*-hexane as eluent or by recrystallization in *n*-hexane to give the [2 + 2] cycloadducts **III-4aa–4an**.

1-(1-Trimethylsilyl-2,3-dihydro-2,3-indolyl)-*o*-carborane **III-4aa**: Colorless crystals. Yield: 86 %. ^{1}H NMR (400 MHz, CD_2Cl_2): δ 7.14 (t, J = 8.0 Hz, 1H), 7.04 (d, J = 7.2 Hz, 1H), 6.74 (t, J = 7.2 Hz, 1H), 6.61 (d, J = 8.0 Hz, 1H) (aromatic C*H*), 5.04 (d, J = 7.2 Hz, 1H), 4.78 (d, J = 7.2 Hz, 1H) (C*H*), 0.33 (s, 9H) (Si(CH_3)$_3$). $^{13}C\{^{1}H\}$ NMR (100 MHz, CD_2Cl_2): δ 156.1, 130.3, 126.0, 125.5, 119.2, 109.8, 87.0 (cage *C*), 82.7 (cage *C*), 71.9, 58.0, −0.8. $^{11}B\{^{1}H\}$ NMR (128 MHz, CD_2Cl_2): δ −4.5 (2B), −8.1 (1B), −11.3 (7B). HRMS (EI) calcd for $C_{13}H_{25}^{11}B_8^{10}B_2NSi^+$: 331.2762. Found 331.2762.

1-(1-Trimethylsilyl-2-phenyl-3-hydro-2,3-indolyl)-*o*-carborane **III-4ab**: Colorless crystals. Yield: 63 %. ^{1}H NMR (400 MHz, CD_2Cl_2): 7.46 (br, 2H), 7.04 (br, 3H), 7.25 (t, J = 7.6 Hz, 1H), 7.15 (d, J = 7.2 Hz, 1H), 6.83 (t, J = 7.6 Hz, 1H), 6.75 (d, J = 8.0 Hz, 1H) (aromatic C*H*), 4.67 (s, 1H) (C*H*), 0.16 (s, 9H) (Si (CH_3)$_3$). $^{13}C\{^{1}H\}$ NMR (100 MHz, CD_2Cl_2): 156.5, 142.0, 130.4, 128.8, 128.3, 127.3, 125.7, 125.4, 119.3, 110.0, 88.2 (cage *C*), 85.9 (cage *C*), 79.7, 66.6, 0.2. $^{11}B\{^{1}H\}$ NMR (128 MHz, CD_2Cl_2): −3.0 (1B), −4.6 (1B), −9.6 (2B), −10.9 (4B), −12.6 (2B). HRMS (EI) calcd for $C_{19}H_{29}^{11}B_8^{10}B_2NSi^+$: 407.3725. Found 407.3726.

1-(1-Trimethylsilyl-2-methyl-3-hydro-2,3-indolyl)-*o*-carborane **III-4ac**: White solid. Yield: 69 %. ^{1}H NMR (400 MHz, CD_2Cl_2): δ 6.99 (d, J = 8.0 Hz, 1H), 6.88 (d, J = 7.6 Hz, 1H), 6.67 (d, J = 7.6 Hz, 1H), 6.54 (d, J = 8.0 Hz, 1H) (aromatic C*H*), 4.67 (s, 1H) (C*H*), 2.27 (s, 3H) (CH_3), 0.34 (s, 9H) (Si(CH_3)$_3$). $^{13}C\{^{1}H\}$ NMR (100 MHz, CD_2Cl_2): δ 153.8, 130.8, 128.7, 123.0, 119.3, 109.5, 87.1, 85.4 (cage *C*), 82.8 (cage *C*), 72.1, 20.8, −0.7. $^{11}B\{^{1}H\}$ NMR (128 MHz, CD_2Cl_2): δ −4.4 (2B), −8.2 (1B), −11.4 (7B). HRMS (EI) calcd for $C_{14}H_{27}^{11}B_8^{10}B_2NSi^+$: 345.3007. Found 345.3010.

1-(1-Trimethylsilyl-2-hydro-3-methyl-2,3-indolyl)-*o*-carborane **III-4ad**: Colorless crystals. Yield: 61 %. ^{1}H NMR (400 MHz, CD_2Cl_2): δ 7.14 (t, J = 8.0 Hz, 1H), 7.02 (d, J = 6.8 Hz, 1H), 6.78 (t, J = 6.8 Hz, 1H), 6.61 (d, J = 8.0 Hz, 1H) (aromatic C*H*), 4.73 (s, 1H) (C*H*), 1.76 (s, 3H) (CH_3), 0.35 (s, 9H) (Si(CH_3)$_3$). $^{13}C\{^{1}H\}$ NMR (100 MHz, CD_2Cl_2): δ 132.5, 130.8, 130.1, 123.8, 119.4, 109.7, 87.1 (cage *C*), 83.5 (cage *C*), 77.6, 68.6, 23.7, −0.6. $^{11}B\{^{1}H\}$ NMR (128 MHz, CD_2Cl_2): δ −4.2 (2B), −10.6 (1B), −11.8 (7B). HRMS (EI) calcd for $C_{14}H_{27}^{11}B_8^{10}B_2NSi^+$: 345.3007. Found 345.3005.

1-(1-Trimethylsilyl-2,3-dimethyl-2,3-indolyl)-*o*-carborane **III-4ae**: White solid. Yield: 78 %. ^{1}H NMR (400 MHz, CD_2Cl_2): δ 7.50 (d, J = 8.0 Hz, 1H), 7.42 (m, 1H), 7.37 (m, 1H), 7.42 (t, J = 8.0 Hz, 1H) (aromatic C*H*), 2.42 (s, 3H), 2.22 (s, 3H) (CH_3), 0.35 (s, 9H) (Si(CH_3)$_3$). $^{13}C\{^{1}H\}$ NMR (100 MHz, CD_2Cl_2): δ 153.8, 130.8, 128.7, 126.1, 119.3, 109.5, 87.1 (cage *C*), 82.8 (cage *C*), 72.1, 65.2, 24.5, 20.8, −0.7. $^{11}B\{^{1}H\}$ NMR (128 MHz, CD_2Cl_2): δ −3.9 (2B), −7.7 (1B), −11.2 (7B). HRMS (EI) calcd for $C_{15}H_{29}^{11}B_8^{10}B_2NSi^+$: 359.3294. Found 359.3291.

1-(1-Trimethylsilyl-2,3-dihydro-4-methyl-2,3-indolyl)-*o*-carborane **III-4af**: White solid. Yield: 86 %. ^{1}H NMR (400 MHz, CD_2Cl_2): δ 7.05 (t, J = 8.0 Hz, 1H), 7.56 (d, J = 8.0 Hz, 1H), 6.45 (d, J = 6.8 Hz, 1H) (aromatic C*H*), 5.05 (d, J = 7.2 Hz, 1H), 4.80 (d, J = 7.2 Hz, 1H) (C*H*), 2.14 (s, 3H) (CH_3), 0.33 (s, 9H) ($Si(CH_3)_3$). $^{13}C\{^1H\}$ NMR (100 MHz, CD_2Cl_2): δ 155.9, 135.9, 130.5, 125.0, 120.4, 107.3, 87.0 (cage *C*), 82.6 (cage *C*), 71.8, 57.5, 18.8, −0.7. $^{11}B\{^1H\}$ NMR (128 MHz, CD_2Cl_2): δ −3.9 (2B), −7.7 (1B), −11.2 (7B). HRMS (EI) calcd for $C_{14}H_{27}^{11}B_8^{10}B_2NSi^+$: 345.3007. Found 345.3005.

1-(1-Trimethylsilyl-2,3-dihydro-5-methyl-2,3-indolyl)-*o*-carborane **III-4ag**: White solid. Yield: 82 %. ^{1}H NMR (400 MHz, CD_2Cl_2): δ 6.99 (d, J = 8.0 Hz, 1H), 6.89 (s, 1H), 6.54 (d, J = 8.0 Hz, 1H) (aromatic C*H*), 5.04 (d, J = 7.2 Hz, 1H), 4.75 (d, J = 7.2 Hz, 1H) (C*H*), 2.27 (s, 3H) (CH_3), 0.34 (s, 9H) ($Si(CH_3)_3$). $^{13}C\{^1H\}$ NMR (100 MHz, CD_2Cl_2): δ 153.8, 130.8, 128.8, 126.1, 126.1, 109.5, 87.1 (cage *C*), 82.8 (cage *C*), 72.1, 58.2, 20.8, −0.7. $^{11}B\{^1H\}$ NMR (128 MHz, CD_2Cl_2): δ −4.4 (2B), −8.2 (1B), −11.4 (7B). HRMS (EI) calcd for $C_{14}H_{27}^{11}B_8^{10}B_2NSi^+$: 345.3007. Found 345.3008.

1-(1-Trimethylsilyl-2,3-dihydro-5-methoxy-2,3-indolyl)-*o*-carborane **III-4ah**: White solid. Yield: 85 %. ^{1}H NMR (400 MHz, CD_2Cl_2): δ 6.71 (d, J = 8.8 Hz, 1H), 6.65 (s, 1H), 6.51 (d, J = 8.8 Hz, 1H) (aromatic C*H*), 5.03 (d, J = 7.2 Hz, 1H), 4.75 (d, J = 7.2 Hz, 1H) (C*H*), 3.72 (s, 3H) (CH_3), 0.30 (s, 9H) ($Si(CH_3)_3$). $^{13}C\{^1H\}$ NMR (100 MHz, CD_2Cl_2): δ 146.1, 133.3, 130.2, 126.0, 125.5, 122.2, 83.3 (cage *C*), 82.4 (cage *C*), 63.5, 22.2, 0.5. $^{11}B\{^1H\}$ NMR (128 MHz, CD_2Cl_2): δ −3.9 (2B), −7.6 (1B), −11.2 (7B). HRMS (EI) calcd for $C_{14}H_{27}^{11}B_8^{10}B_2NOSi^+$: 361.3022. Found 361.3018.

1-(1-Trimethylsilyl-2,3-dihydro-5-fluoro-2,3-indolyl)-*o*-carborane **III-4ai**: White solid. Yield: 88 %. ^{1}H NMR (400 MHz, CD_2Cl_2): δ 6.87 (m, 1H), 6.79 (m, 1H), 6.50 (m, 1H) (aromatic C*H*), 5.06 (d, J = 7.2 Hz, 1H), 4.76 (d, J = 7.2 Hz, 1H) (C*H*), 0.31 (s, 9H) ($Si(CH_3)_3$). $^{13}C\{^1H\}$ NMR (100 MHz, CD_2Cl_2): δ 156.8 ($^1J_{C\text{–}F}$ = 235.0 Hz), 152.3 ($^3J_{C\text{–}F}$ = 1.0 Hz), 127.6 ($^3J_{C\text{–}F}$ = 8.0 Hz), 116.5 ($^2J_{C\text{–}F}$ = 23.0 Hz), 112.8 ($^2J_{C\text{–}F}$ = 24.0 Hz), 109.8 ($^3J_{C\text{–}F}$ = 8.0 Hz), 86.8 (cage *C*), 82.0 (cage *C*), 72.5, 57.7 ($^3J_{C\text{–}F}$ = 1.0 Hz), −0.9. $^{11}B\{^1H\}$ NMR (128 MHz, CD_2Cl_2): δ −4.2 (2B), −8.2 (1B), −11.2 (4B), −11.7 (3B). HRMS (EI) calcd for $C_{13}H_{24}^{11}B_8^{10}B_2NSiF^+$: 349.2667. Found 349.2661.

1-(1-Trimethylsilyl-2,3-dihydro-5-bromo-2,3-indolyl)-*o*-carborane **III-4aj**: White solid. Yield: 86 %. ^{1}H NMR (400 MHz, CD_2Cl_2): δ 7.24 (d, J = 8.4 Hz, 1H), 7.15 (s, 1H), 6.50 (d, J = 8.4 Hz, 1H) (aromatic C*H*), 5.05 (d, J = 7.2 Hz, 1H), 4.74 (d, J = 7.2 Hz, 1H) (C*H*), 0.33 (s, 9H) ($Si(CH_3)_3$). $^{13}C\{^1H\}$ NMR (100 MHz, CD_2Cl_2): δ 155.4, 133.0, 128.3, 128.2, 111.1, 110.5, 86.6 (cage *C*), 82.0 (cage *C*), 72.2, 57.4, −0.9. $^{11}B\{^1H\}$ NMR (128 MHz, CD_2Cl_2): δ −4.0 (2B), −7.9 (1B), −11.0 (7B). HRMS (EI) calcd for $C_{13}H_{24}^{11}B_8^{10}B_2NSiBr^+$: 410.1724. Found 410.1726.

1-(1-Trimethylsilyl-2,3-dihydro-6-isopropyl-2,3-indolyl)-*o*-carborane **III-4ak**: White solid. Yield: 84 %. ^{1}H NMR (400 MHz, CD_2Cl_2): δ 6.98 (d, J = 7.6 Hz, 1H), 6.65 (dd, J = 7.6, 0.8 Hz, 1H), 6.50 (s, 1H) (aromatic C*H*), 5.04

(d, $J = 7.2$ Hz, 1H), 4.75 (d, $J = 7.2$ Hz, 1H) (CH), 2.87 (m, 1H) (CH), 1.26 (d, $J = 7.2$ Hz, 6H) (CH_3), 0.37 (s, 9H) (Si(CH_3)$_3$). $^{13}C\{^1H\}$ NMR (100 MHz, CD_2Cl_2): δ 156.4, 151.9, 125.2, 123.5, 117.5, 108.1, 87.0 (cage C), 83.0 (cage C), 72.2, 57.9, 34.8, 24.3, 24.3, −0.6. $^{11}B\{^1H\}$ NMR (128 MHz, CD_2Cl_2): δ −4.6 (2B), −8.4 (1B), −11.5 (7B). HRMS (EI) calcd for $C_{16}H_{31}^{11}B_8^{10}B_2NSi^+$: 373.3593. Found 373.3591.

1-(1-Trimethylsilyl-2,3-dihydro-5-chloro-2,3-indolyl)-*o*-carborane **III-4al**: White solid. Yield: 92 %. ^1H NMR (400 MHz, CD_2Cl_2): δ 6.85 (m, 1H), 6.78 (m, 1H), 6.49 (s, 1H) (aromatic CH), 5.05 (d, $J = 7.2$ Hz, 1H), 4.75 (d, $J = 7.2$ Hz, 1H) (CH), 2.87 (m, 1H) (CH), 1.26 (d, $J = 7.2$ Hz, 6H) (CH_3), 0.30 (s, 9H) (Si (CH_3)$_3$). $^{13}C\{^1H\}$ NMR (100 MHz, CD_2Cl_2): δ 155.8, 131.0, 126.4, 125.2, 118.8, 109.3, 86.6 (cage C), 82.2 (cage C), 71.4, 57.2, −0.6. $^{11}B\{^1H\}$ NMR (128 MHz, CD_2Cl_2): δ −4.2 (2B), −8.8 (1B), −11.7 (7B). HRMS (EI) calcd for $C_{13}H_{24}^{11}B_8^{10}B_2NClSi^+$: 365.7216. Found 365.7219.

1-(1-Trimethylsilyl-2,3-dihydro-7-methyl-2,3-indolyl)-*o*-carborane **III-4am**: Colorless crystals. Yield: 86 %. ^1H NMR (400 MHz, CD_2Cl_2): δ 7.00 (d, $J = 7.6$ Hz, 1H), 6.89 (d, $J = 7.2$ Hz, 1H), 6.72 (t, $J = 7.2$ Hz, 1H) (aromatic CH), 5.08 (d, $J = 7.2$ Hz, 1H), 4.77 (d, $J = 7.2$ Hz, 1H) (CH), 2.40 (s, 1H) (CH_3), 0.40 (s, 9H) (Si(CH_3)$_3$). $^{13}C\{^1H\}$ NMR (100 MHz, CD_2Cl_2): δ 154.7, 133.8, 127.1, 123.0, 120.2, 119.5, 87.5 (cage C), 83.1 (cage C), 72.0, 57.7, 21.5, 3.3. $^{11}B\{^1H\}$ NMR (128 MHz, CD_2Cl_2): δ −3.9 (2B), −7.8 (1B), −11.0 (7B). HRMS (EI) calcd for $C_{14}H_{27}^{11}B_8^{10}B_2NSi^+$: 345.3007. Found 345.3008.

1-(Trimethylsilyl-2,3-dihydro-7-bromo-2,3-indolyl)-*o*-carborane **III-4an**: White solid. Yield: 84 %. ^1H NMR (400 MHz, CD_2Cl_2): δ 7.36 (d, $J = 7.6$ Hz, 1H), 6.98 (d, $J = 7.2$ Hz, 1H), 6.66 (t, $J = 7.6$ Hz, 1H) (aromatic CH), 5.12 (d, $J = 7.2$ Hz, 1H), 4.78 (d, $J = 7.2$ Hz, 1H) (CH), 0.50 (s, 9H) (Si(CH_3)$_3$). $^{13}C\{^1H\}$ NMR (100 MHz, CD_2Cl_2): δ 154.6, 135.9, 129.3, 124.4, 121.3, 101.8, 86.9 (cage C), 82.2 (cage C), 71.9, 57.4, 4.4. $^{11}B\{^1H\}$ NMR (128 MHz, CD_2Cl_2): δ −4.0 (2B), −8.2 (1B), −11.3 (7B). HRMS (EI) calcd for $C_{13}H_{24}^{11}B_8^{10}B_2NSiBr^+$: 410.1724. Found 410.1728.

General Procedure for Effect of *N*-substituents on the Reaction of *N*-protected Indoles with *o*-Carboryne. A Representative Procedure To an in situ-prepared suspension of 1-Br-2-Li-*o*-$C_2B_{10}H_{10}$ (1.0 mmol) in *n*-hexane (10 mL), was added 1.1 equiv of *N*-protected indoles **III-3** at room temperature. After stirring for 15 min at room temperature, the resulting mixture was heated at 60 °C for 6 h and then quenched by wet *n*-hexane. The inorganic salt was removed by filtration and washed with *n*-hexane (5 mL × 3). After removal of solvents in vacuo, the residue was examined by ^1H NMR spectroscopy and then purified by flash column chromatography on silica gel (230–400 mesh) using *n*-hexane/EtOAc as eluent or by recrystallization in *n*-hexane to give the [2 + 2] cycloadducts **III-4** or/and the sp^2 C–H insertion products **III-5**.

1-(1-*tert*-Butyldimethylsilyl-2,3-dihydro-2,3-indolyl)-*o*-carborane **III-4b**: White solid. Yield: 33 %. ^1H NMR (400 MHz, CD_2Cl_2): δ 7.93 (t, $J = 4.0$ Hz, 1H), 7.55

(d, J = 4.0 Hz, 1H), 7.22 (m, 2H) (aromatic C*H*), 5.04 (d, J = 7.2 Hz, 1H), 4.78 (d, J = 7.2 Hz, 1H) (C*H*), 0.94 (s, 9H), 0.65 (s, 6H) (CH_3). $^{13}C\{^1H\}$ NMR (100 MHz, CD_2Cl_2): δ 155.7, 131.0, 126.4, 125.2, 118.8, 109.3, 86.6 (cage *C*), 82.2 (cage *C*), 71.4, 57.2, 26.2, 19.7, −0.8. $^{11}B\{^1H\}$ NMR (128 MHz, CD_2Cl_2): δ −3.9 (2B), −7.7 (1B), −11.2 (7B). HRMS (EI) calcd for $C_{16}H_{31}^{11}B_8^{10}B_2NSi^+$: 373.3561. Found 373.3563.

1-(1-*tert*-Butyldimethylsilyl-3-indolyl)-*o*-carborane **III-5b**: White solid. Yield: 42 %. ^1H NMR (400 MHz, CD_2Cl_2): δ 7.93 (m, 1H), 7.55 (m, 1H), 7.42 (s, 1H), 7.22 (m, 2H) (aromatic C*H*), 4.06 (br, 1H) (cage C*H*), 0.94 (s, 9H), 0.65 (s, 6H) (CH_3). $^{13}C\{^1H\}$ NMR (100 MHz, CD_2Cl_2): δ 141.7, 133.8, 128.5, 122.9, 121.5, 120.0, 115.1, 112.1, 74.1 (cage *C*), 64.7 (cage *C*), 26.4, 19.6, −3.8. $^{11}B\{^1H\}$ NMR (128 MHz, CD_2Cl_2): δ −2.2 (1B), −5.2 (1B), −9.8 (2B), −10.3 (2B), −12.2 (2B), −13.1 (2B). HRMS (EI) calcd for $C_{16}H_{31}^{11}B_8^{10}B_2NSi^+$: 373.3561. Found 373.3567.

1-(1-Triisopropylsilyl-3-indolyl)-*o*-carborane **III-5c**: White solid. Yield: 83 %. ^1H NMR (400 MHz, CD_2Cl_2): δ 7.92 (t, J = 7.2 Hz, 1H), 7.51 (m, 2H), 7.21 (m, 2H) (aromatic C*H*), 4.06 (br, 1H) (cage C*H*), 1.71 (m, 3H) (C*H*), 1.15 (d, J = 7.6 Hz, 18H) (CH_3). $^{13}C\{^1H\}$ NMR (100 MHz, CD_2Cl_2): δ 141.5, 134.3, 128.5, 122.9, 121.3, 119.9, 115.1, 112.0, 74.3 (cage *C*), 64.7 (cage *C*), 18.2, 13.1. $^{11}B\{^1H\}$ NMR (128 MHz, CD_2Cl_2): δ −2.4 (1B), −4.9 (1B), −9.7 (2B), −10.3 (2B), −12.1 (2B), −13.1 (2B). HRMS (EI) calcd for $C_{19}H_{37}^{11}B_8^{10}B_2NSi^+$: 415.4359. Found 415.4363.

1-(1-*H*-3-Indolyl)-*o*-carborane **III-5d**: Colorless oil. Yield: 30 %. ^1H NMR (400 MHz, CD_2Cl_2): δ 8.58 (br, 1H) (N*H*), 7.93 (d, J = 7.6 Hz, 1H), 7.48 (s, 1H), 7.42 (d, J = 7.6 Hz, 1H), 7.23 (m, 2H) (aromatic C*H*), 4.03 (br, 1H) (cage C*H*). $^{13}C\{^1H\}$ NMR (100 MHz, CD_2Cl_2): δ 136.3, 127.1, 125.4, 123.5, 121.6, 120.0, 112.3, 110.3, 74.0 (cage *C*), 64.9 (cage *C*). $^{11}B\{^1H\}$ NMR (128 MHz, CD_2Cl_2): δ −2.5 (1B), −5.2 (1B), −9.8 (2B), −10.3 (2B), −12.1 (2B), −13.2 (2B). HRMS (EI) calcd for $C_{10}H_{17}^{11}B_8^{10}B_2N^+$: 259.2365. Found 259.2363.

1-(1-Methyl-3-indolyl)-*o*-carborane **III-5ea**: Colorless crystals. Yield: 85 %. ^1H NMR (400 MHz, CD_2Cl_2): δ 7.91 (d, J = 8.0 Hz, 1H), 7.34 (m, 2H), 7.27 (t, J = 7.6 Hz, 1H), 7.21 (t, J = 6.8 Hz, 1H) (aromatic C*H*), 4.00 (br, 1H) (cage C*H*), 3.76 (s, 3H) (NCH_3). $^{13}C\{^1H\}$ NMR (100 MHz, CD_2Cl_2): δ 137.2, 131.5, 125.9, 123.0, 121.3, 120.1, 110.5, 108.5, 74.3 (cage *C*), 65.1 (cage *C*), 33.4. ^{11}B $\{^1H\}$ NMR (128 MHz, CD_2Cl_2): δ −2.4 (1B), −5.3 (1B), −9.9 (4B), −12.1 (2B), −13.2 (2B). HRMS (EI) calcd for $C_{11}H_{19}^{11}B_8^{10}B_2N^+$: 273.2521. Found 273.2521.

1-(1-Allyl-3-indolyl)-*o*-carborane **III-5f**: White solid. Yield: 85 %. ^1H NMR (400 MHz, CD_2Cl_2): δ 7.92 (d, J = 8.0 Hz, 1H), 7.36 (s, 1H), 7.32 (d, J = 8.0 Hz, 1H), 7.21 (m, 2H) (aromatic C*H*), 5.96 (m, 1H), 5.26 (dd, J = 10.0, 0.8 Hz, 1H), 5.11 (dd, J = 17.2, 0.8 Hz, 1H) (olefinic C*H*), 4.67 (d, J = 5.6 Hz, 2H) (NCH_2) 3.96 (br, 1H) (cage C*H*). $^{13}C\{^1H\}$ NMR (100 MHz, CD_2Cl_2): δ 136.6, 132.9, 130.5, 126.2, 123.1, 121.4, 120.2, 118.5, 110.9, 109.0, 74.2 (cage *C*), 65.2 (cage *C*), 55.3. $^{11}B\{^1H\}$ NMR (128 MHz, CD_2Cl_2): δ −2.5 (1B), −5.3 (1B), −9.9 (4B), −12.1 (2B), −13.2 (2B). HRMS (EI) calcd for $C_{13}H_{21}^{11}B_8^{10}B_2N^+$: 299.2678. Found 299.2676.

1-(1-Benzyl-3-indolyl)-*o*-carborane **III-5g**: White solid. Yield: 78 %. ^{1}H NMR (400 MHz, CD_2Cl_2): δ 8.00 (m, 1H), 7.48 (s, 1H), 7.31 (m, 4H), 7.20 (m, 2H), 7.12 (d, J = 6.4 Hz, 2H) (aromatic C*H*), 5.29 (s, 2H) (NCH_2) 4.02 (br, 1H) (cage C*H*). $^{13}C\{^{1}H\}$ NMR (100 MHz, CD_2Cl_2): δ 136.7, 136.7, 130.9, 129.3, 128.4, 127.3, 126.3, 123.2, 121.5, 120.3, 111.0, 109.3, 74.0 (cage *C*), 65.1 (cage *C*), 50.8. $^{11}B\{^{1}H\}$ NMR (128 MHz, CD_2Cl_2): δ −2.4 (1B), −5.2 (1B), −9.8 (4B), −12.1 (2B), −13.1 (2B). HRMS (EI) calcd for $C_{17}H_{23}^{11}B_8^{10}B_2N^+$: 349.3264. Found 349.3262.

1-(1-Phenyl-2,3-dihydro-2,3-indolyl)-*o*-carborane **III-4h**: White solid. Yield: 39 %. ^{1}H NMR (400 MHz, CD_2Cl_2): δ 7.38 (t, J = 8.0 Hz, 2H), 7.26 (t, J = 8.0 Hz, 1H), 7.16 (m, 4H), 7.05 (d, J = 8.0 Hz, 1H), 6.89 (t, J = 7.2 Hz, 1H) (aromatic C*H*), 5.21 (d, J = 7.2 Hz, 1H), 4.97 (d, J = 7.2 Hz, 1H) (C*H*). $^{13}C\{^{1}H\}$ NMR (100 MHz, CD_2Cl_2): δ 151.3, 142.1, 130.4, 130.0, 125.9, 125.7, 124.3, 120.9, 119.5, 110.5, 85.7 (cage *C*), 81.9 (cage *C*), 73.7, 57.0. $^{11}B\{^{1}H\}$ NMR (128 MHz, CD_2Cl_2): δ −4.2 (2B), −7.9 (1B), −11.0 (7B). HRMS (EI) calcd for $C_{16}H_{21}^{11}B_8^{10}B_2N^+$: 335.2679. Found 335.2675.

1-(1-Phenyl-3-indolyl)-*o*-carborane **III-5h**: Data were collect from the mixture of **III-4u** and **III-5u**. Yield: 38 %. ^{1}H NMR (400 MHz, CD_2Cl_2): δ 8.04 (m, 1H), 7.61 (s, 1H), 7.58 (m, 2H), 7.48 (m, 4H), 7.28 (m, 2H) (aromatic C*H*), 4.06 (br, 1H) (cage C*H*). $^{13}C\{^{1}H\}$ NMR (100 MHz, CD_2Cl_2): δ 138.6, 136.6, 130.5, 130.3, 128.2, 126.5, 125.3, 123.9, 122.2, 120.4, 111.6, 110.9, 81.9 (cage *C*), 65.0 (cage *C*).

1-[1-(4-Methoxyphenyl)-2,3-dihydro-2,3-indolyl]-*o*-carborane **III-4i**: Data were collect from the mixture of **III-4i** and **III-5i**. Yield: 30 %. ^{1}H NMR (400 MHz, CD_2Cl_2): δ 7.43 (t, J = 8.0 Hz, 2H), 7.29 (t, J = 8.0 Hz, 1H), 7.18 (m, 4H), 7.05 (d, J = 8.0 Hz, 1H) (aromatic C*H*), 5.20 (d, J = 7.2 Hz, 1H), 4.95 (d, J = 7.2 Hz, 1H) (C*H*), 5.38 (s, 3H) (CH_3). $^{13}C\{^{1}H\}$ NMR (100 MHz, CD_2Cl_2): δ 151.6, 143.1, 132.4, 131.3, 126.2, 124.7, 124.1, 120.6, 119.7, 111.5, 85.6 (cage *C*), 81.6 (cage *C*), 73.5, 56.0, 52.3.

1-[1-(4-Methoxyphenyl)-3-indolyl]-*o*-carborane **III-5i**: White solid. Yield: 45 %. ^{1}H NMR (400 MHz, CD_2Cl_2): δ 7.31 (d, J = 4.4 Hz, 1H), 7.23 (m, 4H), 7.21 (m, 2H), 7.14 (m, 2H) (aromatic C*H*), 5.35 (s, 3H) (CH_3), 4.05 (br, 1H) (cage C*H*). $^{13}C\{^{1}H\}$ NMR (100 MHz, CD_2Cl_2): δ 136.7, 136.7, 130.9, 129.3, 128.4, 127.3, 126.3, 123.2, 121.5, 120.3, 111.0, 109.3, 74.0 (cage *C*), 65.1 (cage *C*), 56.3. $^{11}B\{^{1}H\}$ NMR (128 MHz, CD_2Cl_2): δ −2.4 (1B), −5.2 (1B), −9.8 (4B), −12.1 (2B), −13.1 (2B). HRMS (EI) calcd for $C_{17}H_{23}^{11}B_8^{10}B_2NO^+$: 365.3215. Found 365.3218.

1-[1-(4-Fluorophenyl)-2,3-dihydro-2,3-indolyl]-*o*-carborane **III-4j**: Colorless crystals. Yield: 43 %. ^{1}H NMR (400 MHz, CD_2Cl_2): δ 7.25 (t, J = 8.0 Hz, 1H), 7.18 (m, 3H), 7.09 (t, J = 8.8 Hz, 2H), 6.88 (t, J = 8.0 Hz, 2H) (aromatic C*H*), 5.16 (d, J = 7.2 Hz, 1H), 4.98 (d, J = 7.2 Hz, 1H) (C*H*). $^{13}C\{^{1}H\}$ NMR (100 MHz, CD_2Cl_2): δ 159.8 ($^{1}J_{C-F}$ = 242.0 Hz), 151.8, 138.2 ($^{3}J_{C-F}$ = 2.5 Hz), 130.5, 125.9, 125.4, 122.3 ($^{3}J_{C-F}$ = 8.1 Hz), 120.9, 116.8 ($^{2}J_{C-F}$ = 22.6 Hz), 109.9, 85.7 (cage *C*), 81.9 (cage *C*), 73.7, 57.0. $^{11}B\{^{1}H\}$ NMR (128 MHz, CD_2Cl_2): δ −4.3 (2B), −8.0 (1B), −11.0 (7B). HRMS (EI) calcd for $C_{16}H_{20}^{11}B_8^{10}B_2NF^+$: 353.2659. Found 353.2662.

1-[1-(4-Fluorophenyl)-3-indolyl]-*o*-carborane **III-5j**: White solid. Yield: 30 %. ^{1}H NMR (400 MHz, CD_2Cl_2): δ 8.02 (m, 1H), 7.56 (s, 1H), 7.43 (m, 3H), 7.27 (m, 4H) (aromatic C*H*), 4.08 (br, 1H) (cage C*H*). $^{13}C\{^{1}H\}$ NMR (100 MHz, CD_2Cl_2): δ 162.2 (J_{C-F} = 245.7 Hz), 136.9, 134.7 (J_{C-F} = 3.1 Hz), 127.4, 127.4, 124.0, 122.2 (J_{C-F} = 8.1 Hz), 120.4 (J_{C-F} = 22.6 Hz), 117.2, 117.0, 111.4, 111.0, 73.5 (cage *C*), 64.9 (cage *C*). $^{11}B\{^{1}H\}$ NMR (128 MHz, CD_2Cl_2): δ −3.6 (1B), −4.9 (1B), −10.0 (3B), −11.1 (3B), −14.1 (2B). HRMS (EI) calcd for $C_{16}H^{11}_{20}B^{10}_{8}B_2NF^{+}$: 353.2659. Found 353.2662.

General Procedure for sp^2 C–H Insertion Reaction of *N*-methyl Indoles and Heterocycles with *o*-Carboryne. A Representative Procedure To an in situ-prepared suspension of 1-Br-2-Li-*o*-$C_2B_{10}H_{10}$ (1.0 mmol) in *n*-hexane (10 mL) was added 1.1 equiv of indoles **III-3eb–3ek** or heterocycles **III-6–11** at room temperature. After stirring for 15 min at room temperature, the resulting mixture was heated at 60 °C for 6 h and then quenched by wet *n*-hexane. The inorganic salt was removed by filtration and washed with *n*-hexane (5 mL × 3). After removal of solvents in vacuo, the residue was examined by ^{1}H NMR spectroscopy and then subjected to flash column chromatography on silica gel (230–400 mesh) using *n*-hexane/EtOAc as eluent to give the sp^2 C–H insertion products.

1-(1,2-Dimethyl-3-indolyl)-*o*-carborane **III-5eb**: White solid. Yield: 60 %. ^{1}H NMR (400 MHz, CD_2Cl_2): δ 7.83 (d, *J* = 8.0 Hz, 1H), 7.32 (d, *J* = 8.0 Hz, 1H), 7.21 (t, *J* = 7.2 Hz, 1H), 7.14 (t, *J* = 8.0 Hz, 1H) (aromatic C*H*), 4.61 (br, 1H) (cage C*H*), 3.65 (s, 3H) (NCH_3), 2.61 (s, 3H) (CH_3). $^{13}C\{^{1}H\}$ NMR (100 MHz, CD_2Cl_2): δ 138.0, 136.3, 126.0, 122.2, 121.1, 119.5, 110.0, 104.8, 62.3 (cage *C*), 55.3 (cage *C*), 30.3, 13.5. $^{11}B\{^{1}H\}$ NMR (128 MHz, CD_2Cl_2): δ −3.2 (1B), −4.5 (1B), −9.6 (2B), −10.3 (2B), −10.9 (2B), −13.3 (2B). HRMS (EI) calcd for $C_{12}H^{11}_{21}B^{10}_{8}B_2N^{+}$: 287.2678. Found 287.2679.

1-(1,4-Dimethyl-3-indolyl)-*o*-carborane **III-5ec**: White solid. Yield: 91 %. ^{1}H NMR (400 MHz, CD_2Cl_2): δ 7.53 (s, 1H), 7.17 (m, 2H), 7.02 (d, *J* = 6.0 Hz, 1H) (aromatic C*H*), 4.33 (br, 1H) (cage C*H*), 3.74 (s, 3H) (NCH_3), 2.84 (s, 3H) (CH_3). $^{13}C\{^{1}H\}$ NMR (100 MHz, CD_2Cl_2): δ 138.4, 134.2, 129.4, 125.5, 124.3, 123.0, 110.6, 108.7, 75.5 (cage *C*), 63.7 (cage *C*), 33.7, 25.4. $^{11}B\{^{1}H\}$ NMR (128 MHz, CD_2Cl_2): δ −3.3 (1B), −4.7 (1B), −9.5 (2B), −11.3 (2B), −13.7 (4B). HRMS (EI) calcd for $C_{12}H^{11}_{21}B^{10}_{8}B_2N^{+}$: 287.2678. Found 287.2679.

1-(1,5-Dimethyl-3-indolyl)-*o*-carborane **III-5ed**: White solid. Yield: 73 %. ^{1}H NMR (400 MHz, CD_2Cl_2): δ 7.70 (s, 1H), 7.29 (s, 1H), 7.23 (d, *J* = 8.4 Hz, 1H), 7.11 (d, *J* = 8.4 Hz, 1H) (aromatic C*H*), 3.98 (br, 1H) (cage C*H*), 3.72 (s, 3H) (NCH_3), 2.48 (s, 3H) (CH_3). $^{13}C\{^{1}H\}$ NMR (100 MHz, CD_2Cl_2): δ 135.6, 130.9, 126.1, 125.5, 124.7, 119.5, 110.2, 107.9, 74.5 (cage *C*), 65.1 (cage *C*), 33.7, 21.8. $^{11}B\{^{1}H\}$ NMR (128 MHz, CD_2Cl_2): δ −2.6 (1B), −5.4 (1B), −9.9 (4B), −12.1 (2B), −13.2 (2B). HRMS (EI) calcd for $C_{12}H^{11}_{21}B^{10}_{8}B_2N^{+}$: 287.2678. Found 287.2676.

1-(1-Methyl-5-methoxy-3-indolyl)-*o*-carborane **III-5ee**: White solid. Yield: 83 %. ^{1}H NMR (400 MHz, CD_2Cl_2): δ 7.35 (d, J = 2.0 Hz, 1H), 7.30 (s, 1H), 7.23 (d, J = 9.2 Hz, 1H), 7.23 (dd, J = 8.8, 2.4 Hz, 1H) (aromatic C*H*), 3.95 (br, 1H) (cage C*H*), 3.86 (s, 3H), 3.72 (s, 3H) (CH_3). $^{13}C\{^{1}H\}$ NMR (100 MHz, CD_2Cl_2): δ 155.5, 132.4, 131.9, 126.4, 113.0, 111.2, 107.8, 102.1, 74.5 (cage *C*), 65.7 (cage *C*), 56.2, 33.6. $^{11}B\{^{1}H\}$ NMR (128 MHz, CD_2Cl_2): δ −2.5 (1B), −5.4 (1B), −9.9 (4B), −12.3 (2B), −13.2 (2B). HRMS (EI) calcd for $C_{12}H_{21}^{11}B_8^{10}B_2NO^+$: 303.2627. Found 303.2624.

1-(1-Methyl-5-bromo-3-indolyl)-*o*-carborane **III-5ef**: White solid. Yield: 85 %. ^{1}H NMR (400 MHz, CD_2Cl_2): δ 8.05 (d, J = 1.6 Hz, 1H), 7.44 (m, 2H), 7.21 (d, J = 8.8 Hz, 1H) (aromatic C*H*), 3.93 (br, 1H) (cage C*H*), 3.75 (s, 3H) (NCH_3). $^{13}C\{^{1}H\}$ NMR (100 MHz, CD_2Cl_2): δ 135.9, 132.7, 127.4, 125.9, 122.5, 114.9, 112.1, 108.1, 73.5 (cage *C*), 65.2 (cage *C*), 33.6. $^{11}B\{^{1}H\}$ NMR (128 MHz, CD_2Cl_2): δ −2.3 (1B), −5.2 (1B), −9.8 (2B), −10.3 (2B), −12.2 (2B), −13.1 (2B). HRMS (EI) calcd for $C_{11}H_{18}^{11}B_8^{10}B_2NBr^+$: 352.1612. Found 352.1613.

1-(1-Methyl-5-fluoro-3-indolyl)-*o*-carborane **III-5eg**: White solid. Yield: 86 %. ^{1}H NMR (400 MHz, CD_2Cl_2): δ 7.61 (dd, J = 10.8, 2.0 Hz, 1H), 7.38 (s, 1H), 7.28 (dd, J = 8.8, 4.4 Hz, 1H), 7.04 (m, 1H) (aromatic C*H*), 3.92 (br, 1H) (cage C*H*), 3.76 (s, 3H) (NCH_3). $^{13}C\{^{1}H\}$ NMR (100 MHz, CD_2Cl_2): δ 159.0 ($J_{C\text{–}F}$ = 234.0 Hz), 133.9, 133.0, 126.3 (d, $J_{C\text{–}F}$ = 10.0 Hz), 111.6 ($J_{C\text{–}F}$ = 10.0 Hz), 111.4 ($J_{C\text{–}F}$ = 6.0 Hz), 105.3 ($J_{C\text{–}F}$ = 26.0 Hz), 73.9 (cage *C*), 65.5 (cage *C*), 33.7. $^{11}B\{^{1}H\}$ NMR (128 MHz, CD_2Cl_2): δ −2.5 (1B), −5.4 (1B), −10.0 (2B), −10.5 (2B), −12.5 (2B), −13.3 (2B). HRMS (EI) calcd for $C_{11}H_{18}^{11}B_8^{10}B_2NF^+$: 291.2098. Found 291.2097.

1-(1-Methyl-6-isopropyl-3-indolyl)-*o*-carborane **III-5eh**: White solid. Yield: 82 %. ^{1}H NMR (400 MHz, CD_2Cl_2): δ 7.84 (d, J = 8.4 Hz, 1H), 7.30 (s, 1H), 7.20 (s, 1H), 7.15 (d, J = 8.4 Hz, 1H) (aromatic C*H*), 3.97 (br, 1H) (cage C*H*), 3.76 (s, 3H) (NCH_3), 3.07 (m, 1H) (C*H*), 1.32 (d, J = 6.8 Hz, 6H) (CH_3). $^{13}C\{^{1}H\}$ NMR (100 MHz, CD_2Cl_2): δ 144.6, 137.5, 131.1, 124.2, 120.8, 119.9, 108.4, 107.7, 74.5 (cage *C*), 65.2 (cage *C*), 34.8, 33.4, 24.6. $^{11}B\{^{1}H\}$ NMR (128 MHz, CD_2Cl_2): δ −2.5 (1B), −5.4 (1B), −9.9 (4B), −12.1 (2B), −13.2 (2B). HRMS (EI) calcd for $C_{10}H_{20}^{11}B_8^{10}B_2N^+$: 315.2992. Found 315.2993.

1-(1-Methyl-6-chloro-3-indolyl)-*o*-carborane **III-5ei**: White solid. Yield: 78 %. ^{1}H NMR (400 MHz, CD_2Cl_2): δ 7.84 (d, J = 8.8 Hz, 1H), 7.34 (m, 2H), 7.18 (dd, J = 8.8, 1.6 Hz, 1H) (aromatic C*H*), 3.96 (br, 1H) (cage C*H*), 3.72 (s, 3H) (NCH_3). $^{13}C\{^{1}H\}$ NMR (100 MHz, CD_2Cl_2): δ 137.7, 132.2, 129.0, 124.6, 121.9, 121.3, 110.5, 109.0, 73.6 (cage *C*), 65.2 (cage *C*), 33.6. $^{11}B\{^{1}H\}$ NMR (128 MHz, CD_2Cl_2): δ −2.5 (1B), −5.2 (1B), −9.9 (2B), −10.4 (2B), −12.3 (2B), −13.2 (2B). HRMS (EI) calcd for $C_{11}H_{18}^{11}B_8^{10}B_2NCl^+$: 308.2105. Found 308.2106.

1-(1,7-Dimethyl-3-indolyl)-*o*-carborane **III-5ej**: White solid. Yield: 73 %. ^{1}H NMR (400 MHz, CD_2Cl_2): δ 7.75 (d, J = 8.4 Hz, 1H), 7.23 (s, 1H), 7.05 (t, J = 7.6 Hz, 1H), 6.96 (d, J = 7.2 Hz, 1H), (aromatic C*H*), 4.01 (s, 3H) (NCH_3), 3.98 (br, 1H) (cage C*H*), 2.73 (s, 3H) (CH_3). $^{13}C\{^{1}H\}$ NMR (100 MHz, CD_2Cl_2): δ 135.9, 133.2, 127.0, 125.6, 122.7, 121.4, 118.0, 108.0, 74.3 (cage *C*), 65.1

(cage *C*), 37.7, 19.9. $^{11}B\{^1H\}$ NMR (128 MHz, CD_2Cl_2): δ −2.6 (1B), −5.3 (1B), −9.9 (4B), −12.1 (2B), −13.2 (2B). HRMS (EI) calcd for $C_{12}H_{21}^{11}B_8^{10}B_2N^+$: 287.2678. Found 287.2673.

1-(1-Methyl-7-bromo-3-indolyl)-*o*-carborane **III-5ek**: White solid. Yield: 88 %. ^1H NMR (400 MHz, CD_2Cl_2): δ 7.90 (d, *J* = 7.6 Hz, 1H), 7.41 (d, *J* = 7.2 Hz, 1H), 7.30 (s, 1H), 7.03 (t, *J* = 7.2 Hz, 1H), (aromatic C*H*), 4.14 (s, 3H) (NCH_3), 3.97 (br, 1H) (cage C*H*). $^{13}C\{^1H\}$ NMR (100 MHz, CD_2Cl_2): δ 134.6, 133.8, 129.0, 128.3, 122.4, 119.7, 108.5, 105.0, 74.5 (cage *C*), 65.1 (cage *C*), 38.0. $^{11}B\{^1H\}$ NMR (128 MHz, CD_2Cl_2): δ −2.4 (1B), −4.9 (1B), −9.7 (2B), −10.3 (2B), −12.1 (2B), −13.1 (2B). HRMS (EI) calcd for $C_{11}H_{18}^{11}B_8^{10}B_2NBr^+$: 352.1612. Found 352.1611.

1-(1-Methyl-2-benzimidazolyl)-*o*-carborane **III-6a**: White solid. Yield: 50 %. ^1H NMR (400 MHz, CD_2Cl_2): δ 7.62 (d, *J* = 8.0 Hz, 1H), 7.37 (d, *J* = 3.6 Hz, 2H), 7.30 (m, 1H) (aromatic C*H*), 5.43 (br, 1H) (cage C*H*), 3.98 (s, 3H) (NCH_3). $^{13}C\{^1H\}$ NMR (100 MHz, CD_2Cl_2): δ 144.3, 140.9, 137.8, 124.6, 123.8, 120.1, 110.6, 68.7 (cage *C*), 62.4 (cage *C*), 32.6. $^{11}B\{^1H\}$ NMR (128 MHz, CD_2Cl_2): δ −2.8 (2B), −8.7 (4B), −13.0 (4B). HRMS (EI) calcd for $C_9H_{25}^{11}B_8^{10}B_2N_2^+$: 274.2473. Found 274.2470.

1-(1-Methyl-2-benzoxazolyl)-*o*-carborane **III-6b**: White solid. Yield: 53 %. ^1H NMR (400 MHz, CD_2Cl_2): δ 7.71 (d, *J* = 7.2 Hz, 1H), 7.58 (d, *J* = 4.4 Hz, 1H), 7.43 (m, 2H) (aromatic C*H*), 4.56 (br, 1H) (cage C*H*). $^{13}C\{^1H\}$ NMR (100 MHz, CD_2Cl_2): δ 156.3, 151.8, 140.8, 127.2, 126.0, 121.0, 111.5, 65.8 (cage *C*), 59.9 (cage *C*). $^{11}B\{^1H\}$ NMR (128 MHz, CD_2Cl_2): δ −2.6 (2B), −8.9 (2B), −11.0 (2B), −11.7 (2B), −13.2 (2B). HRMS (EI) calcd for $C_9H_{15}^{11}B_8^{10}B_2NO^+$: 261.2156. Found 261.2160.

1-(1-Methyl-2-benzothiazolyl)-*o*-carborane **III-6c**: Colorless crystals. Yield: 64 %. ^1H NMR (400 MHz, CD_2Cl_2): δ 7.97 (d, *J* = 8.0 Hz, 1H), 7.89 (d, *J* = 8.0 Hz, 1H), 7.53 (m, 1H), 7.46 (m, 1H) (aromatic C*H*), 4.82 (br, 1H) (cage C*H*). $^{13}C\{^1H\}$ NMR (100 MHz, CD_2Cl_2): δ 161.3, 152.2, 136.4, 127.6, 127.1, 124.0, 122.1, 71.4 (cage *C*), 60.0 (cage *C*). $^{11}B\{^1H\}$ NMR (128 MHz, CD_2Cl_2): δ −3.0 (1B), −3.6 (1B), −8.8 (2B), −10.6 (2B), −11.2 (2B), −13.0 (2B). HRMS (EI) calcd for $C_9H_{15}^{11}B_8^{10}B_2NS^+$: 277.1928. Found 277.1923.

1-(1-Methyl-3-pyrrolyl)-*o*-carborane **III-6da**: Colorless crystals. Yield: 48 %. ^1H NMR (400 MHz, $CDCl_3$): δ 6.72 (br, 1H), 6.43 (br, 1H), 6.14 (m, 1H) (aromatic C*H*), 3.69 (br, 1H) (cage C*H*), 3.60 (s, 3H) (NCH_3). $^{13}C\{^1H\}$ NMR (100 MHz, $CDCl_3$): δ 122.5, 122.2, 118.9, 109.4, 73.4 (cage *C*), 64.0 (cage *C*), 36.6. $^{11}B\{^1H\}$ NMR (128 MHz, $CDCl_3$): δ −2.1 (1B), −6.0 (1B), −10.1 (4B), −12.0 (2B), −13.1 (2B). HRMS (EI) calcd for $C_7H_{17}^{11}B_8^{10}B_2N^+$: 223.2362. Found 223.2363.

1-(1-Methyl-2-pyrrolyl)-*o*-carborane **III-6db**: White solid. Yield: 32 %. ^1H NMR (400 MHz, $CDCl_3$): δ 6.54 (t, *J* = 2.4 Hz, 1H), 6.38 (dd, *J* = 3.6, 1.6 Hz, 1H), 6.00 (t, *J* = 3.6 Hz, 1H) (aromatic C*H*), 3.89 (br, 1H) (cage C*H*), 3.83 (s, 3H) (NCH_3). $^{13}C\{^1H\}$ NMR (100 MHz, $CDCl_3$): δ 127.7, 124.2, 114.5, 107.3, 72.1 (cage *C*), 65.7 (cage *C*), 37.4. $^{11}B\{^1H\}$ NMR (128 MHz, $CDCl_3$): δ −2.1 (1B),

−4.0 (1B), −9.3 (2B), −11.0 (4B), −13.0 (2B). HRMS (EI) calcd for $C_7H^{11}_{17}B^{10}_8B_2N^+$: 223.2362. Found 223.2365.

1-(4,5-Dimethyl-2-thiazolyl)-*o*-carborane **III-6e**: White solid. Yield: 87 %. ^{1}H NMR (400 MHz, CD_2Cl_2): δ 4.59 (br, 1H) (cage C*H*), 2.31 (s, 3H), 2.23(s, 3H) (CH_3). ^{13}C{^{1}H} NMR (100 MHz, CD_2Cl_2): δ 155.5, 149.0, 130.9, 71.9 (cage *C*), 59.9 (cage *C*), 14.6, 11.5. ^{11}B{^{1}H} NMR (128 MHz, CD_2Cl_2): δ −3.3 (1B), −4.5 (1B), −9.0 (2B), −10.7 (2B), −11.2 (2B), −13.2 (2B). HRMS (EI) calcd for $C_7H^{11}_{17}B^{10}_8B_2NS^+$: 255.2083. Found 255.2084.

1-(2-Quinolinyl)-*o*-carborane **III-6f**: Colorless crystals. Yield: 40 %. ^{1}H NMR (400 MHz, $CDCl_3$): δ 8.18 (d, J = 8.4 Hz, 1H), 7.96 (d, J = 8.8 Hz, 1H), 7.84 (d, J = 8.0 Hz, 1H), 7.76 (m, 1H), 7.58 (m, 2H) (aromatic C*H*), 5.25 (br, 1H) (cage C*H*). ^{13}C{^{1}H} NMR (100 MHz, $CDCl_3$): δ 150.6, 146.7, 137.9, 130.8, 129.5, 127.9, 127.9, 127.7, 118.7, 75.7 (cage *C*), 57.1 (cage *C*). ^{11}B{^{1}H} NMR (128 MHz, $CDCl_3$): δ −3.1 (1B), −4.0 (1B), −8.7 (2B), −10.4 (2B), −11.9 (2B), −13.4 (2B). HRMS (EI) calcd for $C_{11}H^{11}_{17}B^{10}_8B_2N^+$: 271.2364. Found 271.2360.

Transformation of the [2 + 2] Cycloadducts

Thermal Rearrangement of [2 + 2] Cyclodduct III-4j A toluene solution (2.0 mL) of [2 + 2] cyclodduct **III-4j** (70.6 mg, 0.2 mmol) was heated for 48 h at 110 °C. After removal of toluene, the resulting mixture was examined by ^{1}H NMR spectroscopy and then subjected to flash column chromatography on silica gel (230–400 mesh) using *n*-hexane/EtOAc as eluent to give sp^2 C–H insertion product **III-5j** (54.4 mg, 77 %).

Acid-Promoted Rearrangement of [2 + 2] Cyclodduct III-4j A dichloromethane solution (2.0 mL) of [2 + 2] cyclodduct **III-4j** (70.6 mg, 0.2 mmol) was stirred for 12 h at 40 °C in the presence of an excess amount of TsOH·H_2O. After removal of the solvent, the resulting mixture was examined by ^{1}H NMR spectroscopy and then subjected to flash column chromatography on silica gel (230–400 mesh) using *n*-hexane/EtOAc as eluent to give sp^2 C–H insertion product **III-5j** (46.6 mg, 66 %).

Silica Gel-Promoted Ring opening of Adduct III-4ah An *n*-hexane solution (3.0 mL) of [2 + 2] cycloadduct **III-4ah** (72.3 mg, 0.2 mmol) was stirred at room temperature for 12 h in the presence of silica gel. After removal of the solvent, the resulting mixture was examined by ^{1}H NMR spectroscopy and then subjected to flash column chromatography on silica gel (230–400 mesh) using *n*-hexane/EtOAc as eluent to give ring-opening product **III-5ah** (75.2 mg, 100 %).

1-(1-Trimethylsilyl-5-methoxy-3-indolyl)-*o*-carborane **III-5ah**: Colorless crystals. Yield: 100 %. ^{1}H NMR (400 MHz, CD_2Cl_2): δ 7.36 (m, 3H), 6.98 (dd, J = 8.8, 2.0 Hz, 1H) (aromatic C*H*), 4.00 (br, 1H) (cage C*H*), 3.86 (s, 3H) (CH_3), 0.55 (s, 9H) (Si(CH_3)$_3$). ^{13}C{^{1}H} NMR (100 MHz, CD_2Cl_2): δ 155.5,

135.7, 133.4, 129.4, 114.6, 112.6, 111.4, 102.6, 74.4 (cage *C*), 65.4 (cage *C*), 56.2, −0.2. $^{11}B\{^1H\}$ NMR (128 MHz, CD_2Cl_2): δ −0.7 (1B), −2.1 (1B), −6.6 (2B), −7.2 (2B), −9.0 (2B), −10.1 (2B). HRMS (EI) calcd for $C_8H_{23}^{11}B_8^{10}B_2NOSi^+$: 361.3022. Found 361.3019.

Desiylation-accelerated Ring opening of Adduct III-4ah An H_2O/*n*-hexane (1:1, v/v) solution (10.0 mL) of [2 + 2] cycloadduct **III-4ah** (72.3 mg, 0.2 mmol) was stirred at room temperature for 0.5 h. After removal of the solvent, the resulting mixture was examined by ^1H NMR spectroscopy and then subjected to flash column chromatography on silica gel (230–400 mesh) using *n*-hexane/EtOAc as eluent to give ring-opening product **III-5dh** (58.1 mg, 100 %).

1-(1-*H*-5-Methoxy-3-indolyl)-*o*-carborane **III-5dh**: Colorless oil. Yield: 100 %. ^1H NMR (400 MHz, CD_2Cl_2): δ 8.71 (br, 1H) (N*H*), 7.41 (d, *J* = 6.4 Hz, 2H), 7.30 (d, *J* = 8.8 Hz, 1H), 6.92 (d, *J* = 8.8 Hz, 1H) (aromatic C*H*), 3.96 (br, 1H) (cage C*H*), 3.89 (s, 3H) (CH_3). $^{13}C\{^1H\}$ NMR (100 MHz, CD_2Cl_2): δ 155.5, 131.5, 127.7, 126.0, 113.5, 113.0, 102.2, 102.6, 74.2 (cage *C*), 65.5 (cage *C*), 56.3. $^{11}B\{^1H\}$ NMR (128 MHz, CD_2Cl_2): δ −2.7 (1B), −5.5 (1B), −10.1 (4B), −12.5 (2B), −13.4 (2B). HRMS (EI) calcd for $C_{11}H_{19}^{11}B_8^{10}B_2NO^+$: 289.2470. Found 289.2474.

Desiylation-promoted Ring opening of Adduct III-4ac An H_2O/*n*-hexane (1:1, v/v) solution (10.0 mL) of [2 + 2] cycloadduct **III-4ac** (69.1 mg, 0.2 mmol) was stirred at room temperature for 0.5 h. After removal of the solvent, the resulting mixture was examined by ^1H NMR spectroscopy and then subjected to flash column chromatography on silica gel (230–400 mesh) using *n*-hexane/EtOAc as eluent to give ring-opening product **III-5dc** (55.0 mg, 100 %).

1-(1-*H*-2-Methyl-3-indolyl)-*o*-carborane **III-5dc**: Colorless oil. Yield: 100 %. ^1H NMR (400 MHz, CD_2Cl_2): δ 8.72 (br, 1H) (N*H*), 7.41 (d, *J* = 6.4 Hz, 2H), 7.30 (d, *J* = 7.6 Hz, 1H), 6.92 (d, *J* = 6.4 Hz, 1H) (aromatic C*H*), 3.96 (br, 1H) (cage C*H*), 3.91 (s, 3H) (CH_3). $^{13}C\{^1H\}$ NMR (100 MHz, CD_2Cl_2): δ 136.5, 131.5, 127.7, 126.0, 113.3, 112.8, 109.7, 102.2, 74.2 (cage *C*), 65.5 (cage *C*), 20.4. $^{11}B\{^1H\}$ NMR (128 MHz, CD_2Cl_2): δ −2.6 (1B), −5.7 (1B), −10.3 (4B), −12.5 (2B), −13.4 (2B). HRMS (EI) calcd for $C_{11}H_{19}^{11}B_8^{10}B_2N^+$: 273.1219. Found 273.1214.

Desiylation-promoted Ring opening of Adduct III-4ae An H_2O/*n*-hexane (1:1, v/v) solution (10.0 mL) of [2 + 2] cycloadduct **III-4ae** (71.8 mg, 0.2 mmol) was stirred at room temperature for 0.5 h. After removal of the solvent, the resulting mixture was examined by ^1H NMR spectroscopy and then subjected to flash column chromatography on silica gel (230–400 mesh) using *n*-hexane/EtOAc as eluent to give ring-opening product **III-8** (60.1 mg, 100 %).

1-(2,3-Dimethyl-3-indoninyl)-*o*-carborane **III-8**: Colorless crystals. Yield: 100 %. ^1H NMR (400 MHz, CD_2Cl_2): 7.50 (d, *J* = 8.0 Hz, 1H), 7.41 (t, *J* = 8.0 Hz, 1H), 6.92 (m, 1H), 7.24 (m, 1H) (aromatic C*H*), 3.53 (br, 1H) (cage C*H*), 2.42 (s, 3H), 1.55 (s, 3H) (CH_3). $^{13}C\{^1H\}$ NMR (100 MHz, CD_2Cl_2): 182.6, 154.5, 141.4, 130.2, 126.2, 123.7, 121.4, 77.2 (cage *C*), 63.1 (cage *C*), 60.3, 24.5, 19.2.

$^{11}B\{^1H\}$ NMR (128 MHz, CD_2Cl_2): −3.2 (1B), −3.8 (1B), −9.3 (2B), −11.4 (2B), −12.0 (2B), −13.6 (2B). HRMS (EI) calcd for $C_{12}H_{21}^{1}{}^{11}B_8^{10}B_2N^+$: 287.2678. Found 287.2681.ew

Control Experiments

Reaction of 1,2-Dilithio-*o*-carborane with *N*-TMS indole under Standard Conditions To an in situ-prepared *n*-hexane suspension of 1,2-Li_2-*o*-$C_2B_{10}H_{10}$ (**III-9**) (1.0 mmol), prepared in situ from the reaction of ^{n}BuLi (1.25 mL, 1.6 M in *n*-hexane, 2.0 mmol) with *o*-carborane (144.2 mg, 1.0 mmol) at 0 °C, was added *N*-TMS indole **III-3aa** (209.1 mg, 1.1 mmol) at room temperature. After stirring for 15 min at room temperature, the resulting mixture was heated at 60 °C for 6 h and then quenched by wet *n*-hexane. After removal of solvent in vacuo, the residue was examined by ^1H NMR spectroscopy and then subjected to flash column chromatography on silica gel (230–400 mesh) using *n*-hexane as eluent to give *o*-carborane (123.2 mg, 85 %).

Reaction of 1,2-Dilithio-*o*-carborane with *N*-methylindole under Standard Conditions To an in situ-prepared *n*-hexane suspension of 1,2-Li_2-*o*-$C_2B_{10}H_{10}$ (**III-9**) (1.0 mmol), prepared in situ from the reaction of ^{n}BuLi (1.25 mL, 1.6 M in *n*-hexane, 2.0 mmol) with *o*-carborane (144.2 mg, 1.0 mmol) at 0 °C, was added *N*-methylindole **III-3ea** (144.3 mg, 1.1 mmol) at room temperature. After stirring for 15 min at room temperature, the resulting mixture was heated at 60 °C for 6 h and then quenched by wet *n*-hexane. After removal of solvent in vacuo, the residue was examined by ^1H NMR spectroscopy and then subjected to flash column chromatography on silica gel (230–400 mesh) using *n*-hexane as eluent to give *o*-carborane (118.2 mg, 82 %).

Reaction of 1-Bromo-2-methyl-*o*-carborane/1-Lithio-2-methyl-*o*-carborane with *N*-methylindole under Standard Conditions To a 1:1 molar ratio of 1-bromo-2-methyl-*o*-carborane (**III-10**) (0.5 mmol) and 1-lithio-2-methyl-*o*-carborane (**III-11**) (0.5 mmol) was added *N*-methylindole **III-3ea** (144.3 mg, 1.1 mmol) at room temperature. After stirring for 15 min at room temperature, the resulting mixture was heated at 60 °C for 6 h and then quenched by wet *n*-hexane. After removal of solvents in vacuo, the residue was examined by ^1H NMR spectroscopy and then subjected to flash column chromatography on silica gel (230–400 mesh) using *n*-hexane as eluent. The ^1H, ^{11}B NMR, and GC-MS results showed that most of the crude product was 1-methyl-*o*-carborane (72.8 mg, 46 %) and 1-bromo-2-methyl-*o*-carborane (101.9 mg, 43 %).

1-Methyl-*o*-carborane: ^1H NMR (400 MHz, $CDCl_3$): δ 3.56 (br, 1H) (cage C*H*), 2.03 (s, 3H) (CH_3). $^{11}B\{^1H\}$ NMR (128 MHz, $CDCl_3$): δ −1.9 (1B), −6.9 (1B), −9.3 (2B), −10.8 (2B), −11.5 (2B), −12.9 (2B).

1-Methyl-2-bromo-*o*-carborane: ^{1}H NMR (400 MHz, $CDCl_3$): δ 2.21 (s, 3H) (CH_3). $^{11}B\{^{1}H\}$ NMR (128 MHz, $CDCl_3$): δ −5.3 (3B), −7.6 (2B), −8.5 (2B), −10.1 (3B).

Reaction of 1-Br-2-Li-*o*-carborane with *N*-TMS indole in the Presence of TEMPO To an in situ-prepared suspension of 1-Br-2-Li-*o*-$C_2B_{10}H_{10}$ (1.0 mmol) in *n*-hexane (10 mL) was added TEMPO (171.6 mg, 1.1 mmol) at 0 °C. Then, to this mixture was added 1.1 equiv of *N*-TMS indole **III-3aa** (209.1 mg, 1.1 mmol) at room temperature. After stirring for 15 min at room temperature, the resulting mixture was heated at 60 °C for 6 h and then quenched by wet *n*-hexane. After removal of solvents in vacuo, the residue was examined by ^{1}H NMR spectroscopy and then subjected to flash column chromatography on silica gel (230–400 mesh) using *n*-hexane as eluent to give desired product **III-4aa** (265.1 mg, 80 %). The ^{1}H and GC-MS results showed that most of the crude product was **III-4aa**, with a small amount of the starting material 1-bromo-*o*-carborane coupled with a trace amount of carborane. No any TEMPO-coupled product was detected.

Reaction of 1-Br-2-Li-*o*-carborane with *N*-methylindole in the Presence of TEMPO To an in situ-prepared suspension of 1-Br-2-Li-*o*-$C_2B_{10}H_{10}$ (**1**) (1.0 mmol) in *n*-hexane (10 mL) was added TEMPO (171.6 mg, 1.1 mmol) at 0 °C. Then, to this mixture was added 1.1 equiv of *N*-methylindole **III-3ea** (144.3 mg, 1.1 mmol) at room temperature. After stirring for 15 min at room temperature, the resulting mixture was heated at 60 °C for 6 h and then quenched by wet *n*-hexane. After removal of solvents in vacuo, the residue was examined by ^{1}H NMR spectroscopy and then subjected to flash column chromatography on silica gel (230–400 mesh) using *n*-hexane/EtOAc as eluent to give desired product **III-5ea** (204.9 mg, 75 %). The ^{1}H and GC-MS results showed that most of the crude product was **III-5ea**, with a small amount of the starting material 1-bromo-*o*-carborane coupled with a trace amount of carborane. No any TEMPO-coupled product was detected.

Reaction of 1-Bromo-2-methyl-*o*-carborane/1-Lithio-2-methyl-*o*-carborane with *N*-methylindole under UV light Irradiation (365 nm) To a 1:1 mixture of 1-bromo-2-methyl-*o*-carborane (**III-10**) (0.5 mmol) and 1-lithio-2-methyl-*o*-carborane (**III-11**) (0.5 mmol) was added *N*-methylindole **III-3ea** (144.3 mg, 1.1 mmol) at room temperature. After stirring for 15 min at room temperature, the resulting mixture was heated at 60 °C for 6 h under a handheld UV lamp irradiation (365 nm) and then quenched by wet *n*-hexane. After removal of solvent in vacuo, the residue was examined by ^{1}H NMR spectroscopy and then subjected to flash column chromatography on silica gel (230–400 mesh) using *n*-hexane as eluent. The ^{1}H, ^{11}B NMR, and GC-MS results showed that most of the crude product was 1-methyl-*o*-carborane (70.1 mg, 44 %) and 1-bromo-2-methyl-*o*-carborane (98.0 mg, 41.3 %).

Reaction of 1-Br-2-Li-*o*-carborane with 1,3-Dimethyl-1*H*-indole under Standard Conditions To an in situ-prepared suspension of 1-Br-2-Li-*o*-$C_2B_{10}H_{10}$ (1.0 mmol) in *n*-hexane (10 mL) was added 1,3-dimethyl-1*H*-indole (**III-3el**) (159.7 mg, 1.1 mmol) at room temperature. After stirring for 15 min at room temperature, the resulting mixture was heated at 60 °C for 6 h and then quenched by wet *n*-hexane. After removal of solvents in vacuo, the residue was examined by ^{1}H NMR spectroscopy and then subjected to flash column chromatography on silica gel (230–400 mesh) using *n*-hexane/EtOAc as eluent to give the sp^3 C–H insertion product **III-13** (143.7 mg, 50.0 %) and ene product **III-14** (72.7 mg, 25.3 %). The ene product **III-14** could isomerize to give the rearomatized product **III-5el** in $CDCl_3$ after 6 h.

1-[(3-Methyl-1-indolyl)methyl]-*o*-carborane **III-13**: Colorless oil. Yield: 50 %. ^{1}H NMR (400 MHz, $CDCl_3$): δ 7.21 (t, J = 7.6 Hz, 1H), 7.07 (d, J = 7.6 Hz, 1H), 7.07 (t, J = 7.2 Hz, 1H), 6.45 (d, J = 7.6 Hz, 1H), 5.02 (s, 1H) (aromatic C*H*), 3.34 (br, 1H) (cage C*H*), 2.84 (s, 2H) (NCH_2), 1.54 (s, 3H) (CH_3). ^{13}C{^{1}H} NMR (100 MHz, $CDCl_3$): δ 148.8, 130.5, 130.3, 124.6, 118.4, 107.1, 96.5, 82.7 (cage *C*), 61.9 (cage *C*), 53.2, 30.0, 23.0. ^{11}B{^{1}H} NMR (128 MHz, $CDCl_3$): δ −3.8 (2B), −9.3 (2B), −11.8 (2B), −13.7 (4B). HRMS (EI) calcd for $C_{12}H_{21}^{1}{}^{11}B_8^{10}B_2N^+$: 287.2678. Found 287.2676.

1-(1-Methyl-2,3-dihydro-3-methylene-2-indolyl)-*o*-carborane **III-14**: Colorless oil. Yield: 25 %. ^{1}H NMR (400 MHz, $CDCl_3$): δ 7.28 (d, J = 7.6 Hz, 1H), 7.20 (m, 1H), 6.79 (m, 1H), 6.61 (d, J = 8.0 Hz, 1H) (aromatic C*H*), 5.70 (d, J = 2.0 Hz, 1H), 5.08 (d, J = 1.6 Hz, 1H) (olefin C*H*), 4.39 (t, J = 2.0 Hz, 1H) (C*H*), 4.18 (br, 1H) (cage C*H*), 3.00 (s, 3H) (NCH_3). ^{11}B{^{1}H} NMR (128 MHz, $CDCl_3$): δ −3.8 (1B), −4.4 (1B), −8.6 (2B), −11.6 (2B), −14.1 (4B).

1-(1-Methyl-3-methyl-2-indolyl)-*o*-carborane **III-5el**: White solid. ^{1}H NMR (400 MHz, $CDCl_3$): δ 7.54 (d, J = 8.0 Hz, 1H), 7.30 (m, 2H), 7.17 (m, 1H) (aromatic C*H*), 4.78 (br, 1H) (cage C*H*), 3.96 (s, 3H) (NCH_3), 2.50 (s, 3H) (CH_3). ^{13}C{^{1}H} NMR (100 MHz, $CDCl_3$): δ 138.6, 127.8, 126.3, 124.5, 120.6, 119.3, 112.2, 110.1, 72.7 (cage *C*), 61.1 (cage *C*), 34.2, 12.1. ^{11}B{^{1}H} NMR (128 MHz, $CDCl_3$): δ −2.8 (2B), −9.1 (3B), −10.6 (3B), −13.3 (2B). HRMS (EI) calcd for $C_{12}H_{21}{}^{11}B_8{}^{10}B_2N^+$: 287.2678. Found 287.2675.

Reaction of 1-Br-2-Li-*o*-carborane with 1-TMS-3-methyl-1*H*-indole under Standard Conditions To an in situ-prepared suspension of 1-Br-2-Li-*o*-$C_2B_{10}H_{10}$ (1.0 mmol) in *n*-hexane (10 mL) was added 1-TMS-3-methyl-1*H*-indole (**III-3ad**) (223.6 mg, 1.1 mmol) at room temperature. After stirring for 15 min at room temperature, the resulting mixture was heated at 60 °C for 6 h and then quenched by wet *n*-hexane. After removal of solvents in vacuo, the residue was examined by ^{1}H NMR spectroscopy and then subjected to flash column chromatography on silica gel (230–400 mesh) using *n*-hexane/EtOAc as eluent to give the [2 + 2] cycloaddition product **III-4ad** as major product (221.9 mg, 61.2 %) and an ene product **III-15** as minor product (83.2 mg, 24.4 %).

1-(1-Trimethylsilyl-3-methyl-2-indolyl)-*o*-carborane **III-15**: Data were collect from the mixture of **III-4ad** and **III-15**. Yield: 25 %. ^{1}H NMR (400 MHz, CD_2Cl_2): δ 7.34 (d, J = 7.2 Hz, 1H), 7.20 (d, J = 7.2 Hz, 1H), 6.86 (t, J = 7.2 Hz, 1H), 6.80 (m, 1H) (aromatic C*H*), 3.83 (br, 1H) (cage C*H*), 1.76 (s, 3H) (CH_3), 0.33 (s, 9H) (Si(CH_3)$_3$). ^{13}C{^{1}H} NMR (100 MHz, CD_2Cl_2): δ 154.3, 153.5, 147.3, 128.9,

121.5, 120.9, 113.6, 108.1, 64.1 (cage *C*), 60.0 (cage *C*), 23.7, −0.4. $^{11}B\{^{1}H\}$ NMR (128 MHz, CD_2Cl_2): δ −2.8 (2B), −9.1 (3B), −10.6 (3B), −13.3 (2B).

Deuterium Experiments

(1) **Reaction of 1-Br-2-Li-*o*-carborane with 3-Deutero-1-methylindole**. To an in situ-prepared suspension of 1-Br-2-Li-*o*-$C_2B_{10}H_{10}$ (1.0 mmol) in *n*-hexane (10 mL) was added 1.1 equiv of 3-deutero-1-methylindole **III-3ea-d$_1$** (91 %-D) at room temperature. After stirring for 15 min at room temperature, the resulting mixture was heated at 60 °C for 6 h, and then, the inorganic salt was removed by filtration and washed with *n*-hexane (5 mL × 3). The filtrate was concentrate to about 3 mL, from which **III-5ea-d$_1$** was precipitated out as colorless crystals (213.1 mg, 78 %). After ^1H NMR analysis, 84 % deuterium incorporation at cage C–H was obtained. However, the deuterium was labile toward air and moisture as monitor by ^1H and ^2H NMR spectra.

(2) **Reaction of 1-Br-2-Li-*o*-carborane with 1-Methylindole in cyclohexane-d$_{12}$**. To an in situ-prepared suspension of 1-Br-2-Li-*o*-$C_2B_{10}H_{10}$ (1.0 mmol) in cyclohexane-d$_{12}$ (10 mL) was added 1.1 equiv of 1-methylindole **III-3ea** at room temperature. After stirring for 15 min at room temperature, the resulting mixture was heated at 60 °C for 6 h and then quenched by wet *n*-hexane. After removal of solvents in vacuo, the residue was examined by ^1H NMR spectroscopy and then subjected to flash column chromatography on silica gel (230–400 mesh) using *n*-hexane/EtOAc as eluent to give desired product **III-5ea** (224.0 mg, 82 %) with 0 % deuterium incorporation at cage H (determined by crude ^1H NMR spectra).

(3) **Reaction of 1-Br-2-Li-*o*-carborane with 1-Methylindole Quenching by D_2O**. To an in situ-prepared suspension of 1-Br-2-Li-*o*-$C_2B_{10}H_{10}$ (1.0 mmol) (**1**) in *n*-hexane (10 mL) was added 1.1 equiv of 1-methylindole **III-3a** at room temperature. After stirring for 15 min at room temperature, the resulting mixture was heated at 60 °C for 6 h and then quenched by D_2O. After removal of solvents in vacuo, the residue was examined by ^1H NMR spectroscopy and then subjected to flash column chromatography on silica gel (230–400 mesh) using *n*-hexane/EtOAc as eluent to give desired product **III-5ea** (229.5 mg, 84 %) with 0 % deuterium incorporation at cage H (determined by crude ^1H NMR spectra).

Debenzylation of III-5g with Aluminum Chloride To a benzene or anisole solution (10 mL) of 1-benzyl-3-carboranyl-1*H*-indole **III-5g** (186.0 mg, 0.5 mmol) was added aluminum chloride (266.7 mg, 2.0 mmol) at 0 °C. After stirring for 3 h, the reaction mixture was quenched by water and extracted by EtOAc (10 mL × 3). The organic layer was combined, washed by 5 % $NaHCO_3$ and brine. Then, it was dried over $MgSO_4$, and evaporated to dryness in vacuo; the residue was examined by ^{1}H NMR spectroscopy and then subjected to flash column chromatography on silica gel (230–400 mesh) using *n*-hexane/EtOAc as eluent to give desired product 3-carboranyl-1*H*-indole **III-5d** (128.1 mg, 98 %).

Reaction of *o*-Carboryne with *N, α*-Diarylnitrones A THF solution (3 mL) of phenyl[*o*-(trimethylsiyl)carboranyl]iodonium acetate (**IV-1**) (0.1 mmol) and *N, α*-diarylnitrones (0.11 mmol) was stirred at 25 °C for 30 min in the presence of CsF (0.2 mmol) to give a red orange solution. After quenching with wet *n*-hexane (2 mL), the mixture was quickly passed through a short pad of silica gel to remove the undissolved inorganic salts and washed with ether (3 mL × 3). After removal of solvents in vacuo, the residue was analyzed by ^{1}H NMR spectroscopy and then subjected to flash column chromatography on silica gel (230–400 mesh) using *n*-hexane/EtOAc as eluent to give the products **IV-4**.

1,2-(4-Phenyl-2,3,4,5-tetrahydro-1,5-benzoxazepin-3,4-yl)-*o*-carborane **IV-4a**: Colorless oil. Yield: 78 %. ^{1}H NMR (400 MHz, CDCl$_3$): δ 7.50 (s, 2H), 7.46 (d, J = 2.8 Hz, 3H), 7.05 (t, J = 7.6 Hz, 1H), 6.99 (t, J = 8.0 Hz, 1H), 6.81 (d, J = 8.0 Hz, 1H), 6.63 (d, J = 8.0 Hz, 1H) (aromatic C*H*), 5.50 (s, 1H) (C*H*), 4.24 (br, 1H) (N*H*). ^{13}C{^{1}H} NMR (100 MHz, CDCl$_3$): δ 144.6, 138.3, 137.3, 130.0, 129.2, 127.9, 127.3, 122.4, 121.1, 119.1, 102.6 (cage *C*), 82.0 (cage *C*), 63.8. ^{11}B{^{1}H} NMR (128 MHz, CDCl$_3$): δ −5.3 (1B), −9.7 (2B), −11.6 (2B), −13.2 (3B), −15.4 (2B). HRMS (EI) calcd for $C_{15}H_{20}^{11}B_8^{10}B_2NO^+$: 340.2602. Found 340.2600.

1,2-[4-(2-Chlorophenyl)-2,3,4,5-tetrahydro-1,5-benzoxazepin-3,4-yl]-*o*-carborane **IV-4b**: Colorless oil. Yield: 67 %. ^{1}H NMR (400 MHz, CDCl$_3$): δ 7.42 (m, 2H), 7.40 (m, 3H), 7.07 (d, J = 1.2 Hz, 1H), 6.99 (dd, J = 22.4, 8.0 Hz, 1H), 6.78 (t, J = 6.8 Hz, 1H), 6.64 (d, J = 8.0 Hz, 1H) (aromatic C*H*), 6.26 (s, 1H) (C*H*), 4.29 (br, 1H) (N*H*). ^{13}C{^{1}H} NMR (100 MHz, CDCl$_3$): δ 131.8, 131.2, 130.5, 130.3, 129.6, 129.3, 128.1, 127.5, 122.6, 122.2, 120.8, 118.9, 108.4, 68.4 (cage *C*), 58.0 (cage *C*). ^{11}B{^{1}H} NMR (128 MHz, CDCl$_3$): δ −4.8 (1B), −9.2 (2B), −11.1 (1B), −12.6 (3B), −14.9 (3B). HRMS (EI) calcd for $C_{15}H_{19}^{11}B_8^{10}B_2NOCl^+$: 374.2213. Found 374.2211.

1,2-[4-(3-Chlorophenyl)-2,3,4,5-tetrahydro-1,5-benzoxazepin-3,4-yl]-*o*-carborane **IV-4c**: Colorless oil. Yield: 76 %. ^{1}H NMR (400 MHz, CDCl$_3$): δ 7.39 (m, 4H), 7.08 (t, J = 7.2 Hz, 1H), 7.04 (d, J = 1.2 Hz, 1H), 6.82 (td, J = 8.0, 1.2 Hz, 1H), 6.65 (dd, J = 8.0, 1.2 Hz, 1H) (aromatic C*H*), 5.46 (s, 1H) (C*H*), 4.17 (br, 1H) (N*H*). ^{13}C{^{1}H} NMR (100 MHz, CDCl$_3$): δ 144.7, 139.1, 138.0, 135.2, 130.5, 130.2, 127.4, 126.2, 124.3, 122.4, 121.5, 119.3, 102.7 (cage *C*), 81.3 (cage *C*), 63.3. ^{11}B{^{1}H} NMR (128 MHz, CDCl$_3$): δ −5.2 (1B), −9.6 (2B), −11.4 (2B), −12.9 (1B), −14.0 (2B), 15.3 (2B). HRMS (EI) calcd for $C_{15}H_{19}^{11}B_8^{10}B_2NOCl^+$: 374.2213. Found 374.2211.

1,2-[4-(4-Chlorophenyl)-2,3,4,5-tetrahydro-1,5-benzoxazepin-3,4-yl]-*o*-carborane **IV-4d**: Colorless oil. Yield: 72 %. ^{1}H NMR (400 MHz, CDCl$_3$): δ 7.45 (m, 4H), 7.04 (m, 1H), 6.99 (d, J = 8.0 Hz, 1H), 6.83 (t, J = 8.0 Hz, 1H), 6.85 (d, J = 8.0 Hz, 1H) (aromatic C*H*), 5.46 (s, 1H) (C*H*), 4.14 (br, 1H) (N*H*). ^{13}C{^{1}H} NMR (100 MHz, CDCl$_3$): δ 144.7, 138.0, 136.0, 135.8, 129.5, 129.2, 127.4, 122.4,

121.5, 119.3, 102.6 (cage *C*), 81.4 (cage *C*), 63.2. $^{11}B\{^{1}H\}$ NMR (128 MHz, $CDCl_3$): δ −5.0 (1B), −8.0 (1B), −9.4 (2B), −11.3 (2B), −12.8 (2B), −15.1 (2B). HRMS (EI) calcd for $C_{15}H_{19}^{11}B_8^{10}B_2NOCl^+$([M-H]$^+$): 373.2133. Found 373.2136.

1,2-[4-(4-Bromophenyl)-2,3,4,5-tetrahydro-1,5-benzoxazepin-3,4-yl]-*o*-carborane **IV-4e**: Colorless oil. Yield: 75 %. ^1H NMR (400 MHz, CD_2Cl_2): δ 7.62 (d, *J* = 8.0 Hz, 2H), 7.42 (d, *J* =8.0 Hz, 2H), 7.08 (t, *J* = 7.6 Hz, 1H), 7.00 (t, *J* = 8.0 Hz, 1H), 6.84 (t, *J* = 7.6 Hz, 1H), 6.70 (t, *J* = 7.6 Hz, 1H) (aromatic C*H*), 5.48 (s, 1H) (C*H*), 4.29 (br, 1H) (N*H*). $^{13}C\{^{1}H\}$ NMR (100 MHz, CD_2Cl_2): δ 138.5, 136.6, 132.6, 131.4, 130.0, 128.7, 127.7, 122.5, 121.6, 119.7, 103.0 (cage *C*), 82.0 (cage *C*), 63.5. $^{11}B\{^{1}H\}$ NMR (128 MHz, CD_2Cl_2): δ −5.0 (1B), −9.3 (2B), −11.3 (2B), −12.8 (3B), −15.1 (2B). HRMS (EI) calcd for $C_{15}H_{19}^{11}B_8^{10}B_2NOBr^+$([M-H]$^+$): 418.1623. Found 418.1625.

1,2-[4-(4-Fluorophenyl)-2,3,4,5-tetrahydro-1,5-benzoxazepin-3,4-yl]-*o*-carborane **IV-4f**: Colorless oil. Yield: 68 %. ^1H NMR (400 MHz, $CDCl_3$): δ 7.62 (dd, *J* = 8.0 Hz, 2H), 7.15 (t, *J* = 8.4 Hz, 2H), 7.05 (t, *J* = 7.2 Hz, 1H), 6.99 (t, *J* = 8.0 Hz, 1H), 6.83 (t, *J* = 7.2 Hz, 1H), 6.65 (d, *J* = 8.0 Hz, 1H) (aromatic C*H*), 5.48 (s, 1H) (C*H*), 4.15 (br, 1H) (N*H*). $^{13}C\{^{1}H\}$ NMR (100 MHz, $CDCl_3$): δ 163.5 (J_{C-F} = 249.0 Hz), 144.7, 138.1, 133.3 (J_{C-F} = 3.0 Hz), 129.7 (J_{C-F} = 8.0 Hz), 127.3, 122.4, 121.4, 119.3, 116.3 (J_{C-F} = 22.0 Hz), 102.6 (cage *C*), 81.7 (cage *C*), 63.1. $^{11}B\{^{1}H\}$ NMR (128 MHz, $CDCl_3$): δ −5.1 (1B), −8.0 (1B), −9.4 (1B), −11.3 (2B), −12.8 (3B), −15.1 (2B). HRMS (EI) calcd for $C_{15}H_{20}^{11}B_8^{10}B_2NOF^+$: 357.2534. Found 357.2533.

1,2-[4-(4-Nitrophenyl)-2,3,4,5-tetrahydro-1,5-benzoxazepin-3,4-yl]-*o*-carborane **IV-4g**: Colorless oil. Yield: 31 %. ^1H NMR (400 MHz, CD_2Cl_2): δ 7.49 (m, 4H), 7.08 (m, 1H), 6.99 (d, *J* = 1.2 Hz, 1H), 6.99 (d, *J* = 1.2 Hz, 1H), 6.83 (m, 1H), 6.70 (dd, *J* = 8.0, 1.2 Hz, 1H) (aromatic C*H*), 5.49 (s, 1H) (C*H*), 4.29 (br, 1H) (N*H*). $^{13}C\{^{1}H\}$ NMR (100 MHz, CD_2Cl_2): δ 145.0, 138.5, 136.1, 136.0, 129.7, 129.6, 127.7, 122.5, 121.6, 119.7, 82.2 (cage *C*), 63.5, another cage *C* was not observed. $^{11}B\{^{1}H\}$ NMR (128 MHz, CD_2Cl_2): δ −5.0 (1B), −9.4 (2B), −11.3 (2B), −12.8 (3B), −15.1 (2B). HRMS (EI) calcd for $C_{15}H_{20}^{11}B_8^{10}B_2N_2O_3^+$: 386.2531. Found 386.2535.

1,2-[4-(4-Methylphenyl)-2,3,4,5-tetrahydro-1,5-benzoxazepin-3,4-yl]-*o*-carborane **IV-4h**: Colorless oil. Yield: 82 %. ^1H NMR (400 MHz, $CDCl_3$): δ 7.39 (d, *J* = 7.6 Hz, 2H), 7.26 (d, *J* = 7.6 Hz, 2H), 7.04 (t, *J* = 8.0 Hz, 1H), 6.98 (d, *J* = 8.0 Hz, 1H), 6.79 (t, *J* = 8.0 Hz, 1H), 6.60 (m, 1H) (aromatic C*H*), 5.47 (s, 1H) (C*H*), 4.22 (br, 1H) (N*H*) 2.40 (s, 3H) (CH_3). $^{13}C\{^{1}H\}$ NMR (100 MHz, $CDCl_3$): δ 134.4, 131.8, 129.8, 127.8, 127.2, 125.7, 122.5, 122.4, 120.9, 119.0, 102.6 (cage *C*), 82.2 (cage *C*), 63.5, 23.1. $^{11}B\{^{1}H\}$ NMR (128 MHz, $CDCl_3$): δ −4.8 (1B), −9.2 (2B), −11.1 (2B), −12.6 (3B), −14.9 (2B). HRMS (EI) calcd for $C_{11}H_{22}^{11}B_8^{10}B_2NO^+$: 353.2785. Found 353.2781.

1,2-[4-(4-Methoxyphenyl)-2,3,4,5-tetrahydro-1,5-benzoxazepin-3,4-yl]-*o*-carborane **IV-4i**: Colorless oil. Yield: 77 %. ^1H NMR (400 MHz, $CDCl_3$): δ 7.46 (br, 4H), 7.05 (m, 1H), 6.99 (d, *J* = 8.0 Hz, 1H), 6.83 (t, *J* = 8.0 Hz, 1H), 6.65

(d, J = 8.0 Hz, 1H) (aromatic C*H*), 5.46 (s, 1H) (C*H*), 4.14 (br, 1H) (N*H*), 3.60 (s, 3H) (CH_3). $^{13}C\{^1H\}$ NMR (100 MHz, $CDCl_3$): δ 144.7, 138.0, 136.0, 135.8, 129.5, 129.2, 127.4, 122.4, 121.5, 119.3, 102.7 (cage *C*), 81.4 (cage *C*), 63.2, 51.8. $^{11}B\{^1H\}$ NMR (128 MHz, $CDCl_3$): δ −5.3 (1B), −9.7 (1B), −11.6 (2B), −13.2 (3B), −15.4 (3B). HRMS (EI) calcd for $C_{16}H_{23}^{11}B_8^{10}B_2NO_2^+$: 371.2786. Found 371.2788.

1,2-[4-(3-Methoxyphenyl)-2,3,4,5-tetrahydro-1,5-benzoxazepin-3,4-yl]-*o*-carborane **IV-4j**: Colorless oil. Yield: 77 %. ^1H NMR (400 MHz, $CDCl_3$): δ 7.50 (s, 1H), 7.46 (m, 3H), 6.87 (d, J = 9.6 Hz, 1H), 6.83 (s, 2H), 6.66 (d, J = 8.6 Hz, 1H) (aromatic C*H*), 5.30 (s, 1H) (C*H*), 4.05 (br, 1H) (N*H*), 3.87 (s, 3H) (CH_3). $^{13}C\{^1H\}$ NMR (100 MHz, $CDCl_3$): δ 143.1, 140.2, 139.1, 136.3, 134.8, 132.4, 128.9, 127.3, 125.4, 123.4, 123.0, 120.3, 103.7 (cage *C*), 82.3 (cage *C*), 64.3, 58.5. $^{11}B\{^1H\}$ NMR (128 MHz, $CDCl_3$): δ −4.8 (1B), −9.2 (1B), −11.1 (2B), −12.6 (3B), −14.9 (3B). HRMS (EI) calcd for $C_{16}H_{23}^{11}B_8^{10}B_2NO_2^+$: 371.2786. Found 371.2788.

1,2-[4-(2-Methoxyphenyl)-2,3,4,5-tetrahydro-1,5-benzoxazepin-3,4-yl]-*o*-carborane **IV-4k**: Colorless oil. Yield: 71 %. ^1H NMR (400 MHz, $CDCl_3$): δ 7.78 (d, J = 13.2 Hz, 2H), 7.26 (m, 2H), 7.13 (m, 1H), 7.08 (m, 1H), 6.79 (t, J = 8.0 Hz, 1H), 6.61 (d, J = 8.8 Hz, 1H) (aromatic C*H*), 5.47 (s, 1H) (C*H*), 4.29 (br, 1H) (N*H*), 4.15 (s, 3H) (CH_3). $^{13}C\{^1H\}$ NMR (100 MHz, $CDCl_3$): δ 144.7, 141.5, 137.5, 136.1, 133.8, 130.5, 130.2, 126.3, 124.7, 123.8, 122.5, 121.6, 101.9 (cage *C*), 81.5 (cage *C*), 63.3, 48.8. $^{11}B\{^1H\}$ NMR (128 MHz, $CDCl_3$): δ −4.9 (1B), −9.2 (1B), −11.2 (2B), −12.7 (3B), −14.9 (3B). HRMS (EI) calcd for $C_{16}H_{23}^{11}B_8^{10}B_2NO_2^+$: 371.2786. Found 371.2788.

1,2-[4-(2-Bromo-4-chlorophenyl)-2,3,4,5-tetrahydro-1,5-benzoxazepin-3,4-yl]-*o*-carborane **IV-4l**: Colorless oil. Yield: 63 %. ^1H NMR (400 MHz, $CDCl_3$): δ 7.56 (d, J = 7.6 Hz, 1H), 7.25 (m, 2H), 7.03 (t, J = 8.8 Hz, 1H), 6.99 (d, J = 8.0 Hz, 1H), 6.77 (m, 1H), 6.61 (d, J = 8.8 Hz, 1H) (aromatic C*H*), 5.60 (s, 1H) (C*H*), 4.63 (br, 1H) (N*H*). $^{13}C\{^1H\}$ NMR (100 MHz, $CDCl_3$): δ 144.8, 135.2, 133.8, 131.4, 128.0, 127.4, 126.3, 125.7, 124.0, 122.8, 121.9, 120.7, 102.7 (cage *C*), 81.3 (cage *C*), 66.6. $^{11}B\{^1H\}$ NMR (128 MHz, $CDCl_3$): δ −4.9 (1B), −9.2 (1B), −11.2 (2B), −12.7 (3B), −15.0 (3B). HRMS (EI) calcd for $C_{15}H_{19}^{11}B_8^{10}B_2NOClBr^+$: 453.1396. Found 453.1399.

1,2-(4-Phenyl-8-methoxy-2,3,4,5-tetrahydro-1,5-benzoxazepin-3,4-yl)-*o*-carborane **IV-4m**: Colorless oil. Yield: 84 %. ^1H NMR (400 MHz, $CDCl_3$): δ 7.50 (d, J = 12.4 Hz, 2H), 7.25 (m, 3H), 6.99 (d, J = 8.0 Hz, 1H), 6.79 (t, J = 8.0 Hz, 1H), 6.61 (d, J = 8.8 Hz, 1H) (aromatic C*H*), 5.39 (s, 1H) (C*H*), 4.22 (br, 1H) (N*H*), 4.04 (s, 3H) (CH_3). $^{13}C\{^1H\}$ NMR (100 MHz, $CDCl_3$): δ 135.2, 131.4, 130.3, 127.4, 124.9, 124.2, 123.4, 122.4, 121.6, 119.3, 102.3 (cage *C*), 81.6 (cage *C*), 62.9, 52.5. $^{11}B\{^1H\}$ NMR (128 MHz, $CDCl_3$): δ −4.8 (1B), −9.1 (2B), −11.1 (2B), −12.7 (1B), −13.4 (2B), −14.9 (2B). HRMS (EI) calcd for $C_{16}H_{23}^{11}B_8^{10}B_2NO_2^+$: 371.2786. Found 371.2788.

1,2-(4-Phenyl-8-methyl-2,3,4,5-tetrahydro-1,5-benzoxazepin-3,4-yl)-*o*-carborane **IV-4n**: Colorless oil. Yield: 78 %. ^1H NMR (400 MHz, $CDCl_3$): δ 7.50

(m, 2H), 7.45 (d, *J* = 2.4 Hz, 3H), 6.87 (d, *J* = 8.0 Hz, 1H), 6.82 (s, 1H), 6.56 (d, *J* = 8.0 Hz, 1H) (aromatic C*H*), 5.38 (s, 1H) (C*H*), 4.08 (br, 1H) (N*H*), 2.27 (s, 3H) (CH_3). $^{13}C\{^1H\}$ NMR (100 MHz, $CDCl_3$): δ 144.9, 137.7, 135.6, 131.6, 129.9, 129.1, 127.9, 127.8, 122.5, 119.4, 102.9 (cage *C*), 82.0 (cage *C*), 64.2, 20.4. $^{11}B\{^1H\}$ NMR (128 MHz, $CDCl_3$): δ −5.3 (1B), −9.7 (2B), −11.6 (2B), −13.2 (3B), −15.4 (2B). HRMS (EI) calcd for $C_{16}H_{23}^{11}B_8^{10}B_2NO^+$: 353.2785. Found 353.2782.

1,2-(4-Phenyl-8-chloro-2,3,4,5-tetrahydro-1,5-benzoxazepin-3,4-yl)-*o*-carborane **IV-4o**: Colorless oil. Yield: 70 %. ^1H NMR (400 MHz, CD_2Cl_2): δ 7.50 (m, 2H), 7.48 (m, 3H), 7.05 (m, 2H), 6.62 (d, *J* = 8.4 Hz, 2H) (aromatic C*H*), 5.52 (s, 1H) (C*H*), 4.44 (br, 1H) (N*H*). $^{13}C\{^1H\}$ NMR (100 MHz, CD_2Cl_2): δ 137.6, 137.1, 130.4, 129.5, 128.2, 127.7, 127.6, 124.9, 122.7, 120.1, 102.4 (cage *C*), 82.4 (cage *C*), 64.8. $^{11}B\{^1H\}$ NMR (128 MHz, CD_2Cl_2): δ −5.1 (1B), −8.0 (1B), −9.4 (1B), −11.3 (2B), −12.8 (3B), −15.1 (2B). HRMS (EI) calcd for $C_{15}H_{19}^{11}B_8^{10}B_2NOCl^+$: 374.2213. Found 374.2211.

1,2-(4-Phenyl-8-trifluoromethyl-2,3,4,5-tetrahydro-1,5-benzoxazepin-3,4-yl)-*o*-carborane **IV-4p**: Colorless oil. Yield: 63 %. ^1H NMR (400 MHz, CD_2Cl_2): δ 7.47 (d, *J* = 8.4 Hz, 1H), 7.42 (d, *J* = 8.4 Hz, 1H), 7.33 (m, 2H), 7.26 (m, 3H), 7.00 (d, *J* = 6.8 Hz, 1H) (aromatic C*H*), 5.96 (s, 1H) (C*H*), 4.18 (br, 1H) (N*H*). $^{13}C\{^1H\}$ NMR (100 MHz, CD_2Cl_2): δ 141.3, 134.3, 131.8 (d, $^3J_{C-F}$ = 9.2 Hz), 129.6 (d, $^2J_{C-F}$ = 15.5 Hz), 128.4 (d, $^3J_{C-F}$ = 9.3 Hz), 126.5 (d, $^3J_{C-F}$ = 3.5 Hz), 125.8 (d, $^2J_{C-F}$ = 3.7 Hz), 124.6, 122.1 (d, $^3J_{C-F}$ = 8.9 Hz), 125.8 (q, $^1J_{C-F}$ = 212.0 Hz), 105.0, 68.3 (cage *C*), 60.7 (cage *C*). $^{11}B\{^1H\}$ NMR (128 MHz, CD_2Cl_2): δ −5.1 (1B), −8.0 (1B), −9.4 (1B), −11.3 (2B), −12.8 (3B), −15.1 (2B). HRMS (EI) calcd for $C_{16}H_{19}^{11}B_8^{10}B_2NOF_3^+$: 408.2476. Found 408.2470.

1,2-(4-Phenyl-8-nitro-2,3,4,5-tetrahydro-1,5-benzoxazepin-3,4-yl)-*o*-carborane **IV-4q**: Colorless oil. Yield: 45 %. ^1H NMR (400 MHz, CD_2Cl_2): δ 7.88 (m, 2H), 7.49 (d, *J* = 2.4 Hz, 1H), 7.45 (d, *J* = 2.4 Hz, 2H), 7.29 (m, 1H), 7.16 (d, *J* = 7.2 Hz, 1H), 6.72 (d, *J* = 8.4 Hz, 1H) (aromatic C*H*), 5.27 (s, 1H) (C*H*), 4.44 (br, 1H) (N*H*). $^{13}C\{^1H\}$ NMR (100 MHz, CD_2Cl_2): δ 137.6, 137.1, 130.4, 129.5, 128.2, 127.7, 127.6, 124.9, 122.7, 120.1, 102.4 (cage *C*), 82.4 (cage *C*), 64.8. $^{11}B\{^1H\}$ NMR (128 MHz, CD_2Cl_2): δ −6.6 (1B), −10.9 (1B), −12.8 (2B), −14.5 (2B), −15.2 (2B), −16.7 (2B). HRMS (EI) calcd for $C_{15}H_{20}^{11}B_8^{10}B_2N_2O_3^+$: 386.2531. Found 386.2534.

1,2-[4-(Pyridin-3-yl)-8-nitro-2,3,4,5-tetrahydro-1,5-benzoxazepin-3,4-yl]-*o*-carborane **IV-4r**: Colorless oil. Yield: 51 %. ^1H NMR (400 MHz, CD_2Cl_2): δ 9.01 (1H), 8.74(s, 1H), 8.28 (d, *J* = 8.0 Hz, 1H), 7.90 (d, *J* = 8.0 Hz, 1H), 7.11 (m, 1H), 7.02 (d, *J* = 8.0 Hz, 1H), 6.88 (m, *J* = 7.2 Hz, 1H), 6.72 (d, *J* = 8.0 Hz, 1H) (aromatic C*H*), 5.54 (s, 1H) (C*H*), 4.31 (br, 1H) (N*H*). $^{13}C\{^1H\}$ NMR (100 MHz, CD_2Cl_2): δ 157.8, 152.4, 151.7, 151.2, 149.7, 135.4, 135.3, 127.8, 126.8, 124.3, 124.2, 62.1, two cages *C* were not observed. $^{11}B\{^1H\}$ NMR (128 MHz, CD_2Cl_2): δ −4.8 (1B), −9.2 (2B), −11.1 (2B), −12.6 (2B), −14.9 (3B). HRMS (EI) calcd for $C_{14}H_{20}^{11}B_8^{10}B_2N_2O^+$: 342.2633. Found 342.2633.

Preparation of 3-Diazonium-*o*-carborane Tetrafluoroborate (V-1)

Method A: Starting from 3-amino-*o*-carborane and $NOBF_4$ To an anhydrous acetonitrile solution (3 mL) of 3-amino-*o*-carborane (477.0 mg, 3.0 mmol) was added $NOBF_4$ (420.5 mg, 3.9 mmol) at −30 °C, and the reaction mixture was stirred at −15 °C for 1.5 h. Dry ether (5 mL) was then added at −10 °C to the reaction mixture from which **V-1** was precipitated out. Compound **V-1** was collected as a white solid (342.6 mg, 45 %) by filtration and washed by dry ether (3 mL × 3).

Method B: Starting from 3-amino-*o*-carborane and NOF generated in situ[2] To an ether solution (10 mL) of boron trifluoride diethyl etherate (0.92 mL, 7.5 mmol) was added 3-amino-*o*-carborane (795.0 mg, 5.0 mmol) at −15 °C. *tert*-Butyl nitrite (0.618 g. 6.0 mmol) in 5 mL of ether was then added dropwise to the rapidly stirred reaction solution over 10 min. The reaction mixture was allowed to stir at 5 °C for 1 h, from which **V-1** was precipitated out. Compound **V-1** was collected as a white solid (888.3 mg, 70 %) by filtration and washed by dry ether (5 mL × 3).

3-Diazonium-*o*-carborane tetrafluoroborate **V-1**: Since **V-1** was slowly decomposed at room temperature, the elemental analysis data and ^{13}C NMR spectrum were unable to be collected. ^{1}H NMR (400 MHz, CD_3CN): δ 5 26 (br, 2H) (cage C*H*). $^{11}B\{^{1}H\}$ NMR (128 MHz, CD_3CN): δ −0.7 (1B, BF4), −0.8 (2B), −5.5 (1B), −8.4 (1B), −11.5 (2B), −12.4 (3B), −15.5 (1B). ^{19}F NMR (376 MHz, CD_3CN): δ −150.2 (s, 4F).

1,3-Dehydro-*o*-carborane Generated by ^{n}BuLi To a suspension of **V-1** (51.6 mg, 0.2 mmol) in benzene (1 mL) was added ^{n}BuLi (0.125 mL, 0.2 mmol, 1.6 M in hexane) at room temperature. The resulting mixture was stirred at room temperature for 10 min and quenched by wet *n*-hexane. After removal of solvents in vacuo, the residue was examined by ^{1}H NMR spectrum and then subjected to flash column chromatography on silica gel (230–400 mesh) using *n*-hexane as eluent to give **V-3a** (16.7 mg, 38 %) as colorless crystals.

1,3-Dehydro-*o*-carborane Generated by Non-nucleophilic Bases To a mixture of **V-1** (51.6 mg, 0.2 mmol) and non-nucleophilic base (0.2 mmol) was added benzene (1 mL) under an atmosphere of dry argon. The resulting mixture was stirred at room temperature for 10 min and quenched by wet *n*-hexane. After removal of solvents in vacuo, the residue was examined by ^{1}H NMR spectrum and then subjected to flash column chromatography on silica gel (230–400 mesh) using *n*-hexane as eluent to give **V-3** as colorless crystals.

Representative Procedure for the Reaction of 1,3-Dehydro-*o*-carborane with Arenes To a mixture of **V-1** (51.6 mg, 0.2 mmol) and LDA (0.2 mmol) was added arene **V-2** (8.0 mmol) under an atmosphere of dry argon. The resulting mixture was stirred at room temperature for 10 min and quenched by wet *n*-hexane. After removal of solvents in vacuo, the residue was examined by ^{1}H NMR spectrum and then subjected to flash column chromatography on silica gel (230–400 mesh) using *n*-hexane as eluent to give the desired products **V-3/V-4**.

1,3-(2,5-Cyclohexadiene-1,4-diyl)-*o*-carborane **V-3a**: Yield: 72 %. Colorless crystals. ^{1}H NMR (400 MHz, $CDCl_3$): δ 6.83 (m, 2H), 6.37 (t, J = 6.4 Hz, 1H), 6.26 (t, J = 6.8 Hz, 1H) (olefinic C*H*), 4.07 (t, J = 6.0 Hz, 1H) (C*H*), 3.85 (br, 1H) (cage C*H*), 3.64 (t, J = 6.0 Hz, 1H) (BC*H*). ^{13}C{^{1}H} NMR (100 MHz, $CDCl_3$): δ 143.3, 142.7, 134.4, 132.6, 82.0 (cage *C*), 70.8 (cage *C*), 45.2, 34.8. ^{11}B{^{1}H} NMR (128 MHz, $CDCl_3$): δ −3.5 (1B), −4.9 (1B), −6.4 (1B), −7.6 (1B), −10.3 (2B), −12.3 (1B), −14.6 (2B), −16.2 (1B). Anal. Calcd for $C_8H_{16}B_{10}$: C, 43.61; H, 7.32. Found: C, 43.65; H, 7.42. HRMS (EI) calcd for $C_8H^{11}_{16}B^{10}_8B_2{}^+$ 220.2254. Found 220.2250.

1,3-(1,2,3,4,5,6-F_6-2,5-cyclohexadiene-1,4-diyl)-*o*-carborane **V-3b**: Yield: 25 %. White solid. Slowly decomposed at room temperature. ^{1}H NMR (400 MHz, $CDCl_3$): δ 4.12 (br, 1H) (cage C*H*). ^{13}C{^{1}H} NMR (100 MHz, $CDCl_3$): δ 72.4 (cage *C*), the olefinic *C*, the bridged *C* connected to the cage B atom, and another cage *C* were not observed. ^{11}B{^{1}H} NMR (128 MHz, $CDCl_3$): δ −0.3 (1B), −3.5 (1B), −4.8 (1B), −7.0 (1B), −9.5 (2B), −12.5 (1B), −14.5 (3B). ^{19}F NMR (376 MHz, CD_3CN): δ −143.2 (s, 1F), −143.5 (s, 1F), −151.6 (s, 1F), −153.4 (s, 1F), −176.6 (s, 1F), −192.1 (s, 1F). HRMS (EI) calcd for $C_8H^{11}_{10}B^{10}_8B_2{}^+F_6$ 328.1689. Found 328.1693.

1,3-(2-Fluoro-2,5-cyclohexadiene-1,4-diyl)-*o*-carborane **V-3ca** or **V-3cd**: Yield: 13 %. White solid. ^{1}H NMR (400 MHz, $CDCl_3$): δ 6.82 (t, J = 6.4 Hz, 1H), 6.46 (t, J = 5.6 Hz, 1H), 5.48 (m, 1H) (olefinic C*H*), 4.01 (br, 1H) (cage C*H*), 3.93 (dd, J = 10.0, 5.6 Hz, 1H) (C*H*), 3.45 (m, 1H) (BC*H*). ^{13}C{^{1}H} NMR (100 MHz, $CDCl_3$): δ 139.9, 136.4, 104.8 (d, $J_{C\text{–}F}$ = 14.0 Hz), 70.7 (cage *C*), 43.5 (d, $J_{C\text{–}F}$ = 6.0 Hz), the olefinic *C* connected to the F atom, the bridged *C* connected to the cage B atom and another cage *C* were not observed. ^{11}B{^{1}H} NMR (128 MHz, $CDCl_3$): δ −2.8 (1B), −4.4 (1B), −5.8 (1B), −8.0 (1B), −9.4 (1B), −10.2 (1B), −12.2 (1B), −14.0 (2B), −15.8 (1B). HRMS (EI) calcd for $C_8H^{11}_{15}B^{10}_8B_2{}^+F$ 238.2160. Found 238.2162.

1,3-(1-Fluoro-2,5-cyclohexadiene-1,4-diyl)-*o*-carborane **V-3cc**: Yield: 22 %. White solid. ^{1}H NMR (400 MHz, $CDCl_3$): δ 6.75 (m, 2H), 6.42 (t, J = 9.2 Hz, 1H), 6.34 (t, J = 9.2 Hz, 1H), (olefinic C*H*), 3.90 (br, 1H) (cage C*H*), 3.45 (m, 1H) (BC*H*). ^{13}C{^{1}H} NMR (100 MHz, $CDCl_3$): δ 140.5 (d, $J_{C\text{–}F}$ = 9.0 Hz), 140.3 (d, $J_{C\text{–}F}$ = 9.0 Hz), 134.4 (d, $J_{C\text{–}F}$ = 27.0 Hz), 104.9 (d, $J_{C\text{–}F}$ = 26.0 Hz), 100.1 (d, $J_{C\text{–}F}$ = 207.0 Hz), 71.0 (cage *C*), the bridged *C* connected to the cage B atom and another cage *C* were not observed. ^{11}B{^{1}H} NMR (128 MHz, $CDCl_3$): δ −2.2 (1B), −4.7 (1B), −6.1 (1B), −8.4 (1B), −9.4 (1B), −10.4 (1B), −13.1 (1B), −14.9 (2B), −16.2 (1B). HRMS (EI) calcd for $C_8H^{11}_{15}B^{10}_8B_2{}^+F$ 238.2160. Found 238.2167.

1,3-(2,3-Dichloro-2,5-cyclohexadiene-1,4-diyl)-*o*-carborane **V-3ea** or **V-3ed**: Yield: 19 %. White solid. ^{1}H NMR (400 MHz, $CDCl_3$): δ 6.87 (t, J = 6.8 Hz, 1H), 6.44 (d, J = 6 Hz, 1H) (olefinic C*H*), 4.11 (d, J = 6.4 Hz, 1H) (C*H*), 3.68 (d, J = 6.8 Hz, 1H) (BC*H*), 3.51 (br, 1H) (cage C*H*). ^{13}C{^{1}H} NMR (100 MHz, $CDCl_3$): δ 141.0, 140.8, 133.3, 129.8, 81.6 (cage *C*), 70.2 (cage *C*), 53.0, 42.4.

$^{11}B\{^1H\}$ NMR (128 MHz, $CDCl_3$): δ −3.0 (1B), −4.9 (1B), −6.2 (1B), −7.6 (1B), −9.5 (1B), −10.6 (1B), −12.8 (1B), −14.0 (2B), −14.5 (1B), −16.0 (1B). HRMS (EI) calcd for $C_8H_{14}^{11}B_8^{10}B_2{}^+Cl_2$ 288.1475. Found 288.1471.

1,3-(1,2-Dichloro-2,5-cyclohexadiene-1,4-diyl)-*o*-carborane **V-3eb** or **V-3ec**: Yield: 23 %. White solid. ^1H NMR (400 MHz, $CDCl_3$): δ 6.81 (m, 2H), 6.35 (d, J = 6.8 Hz, 1H) (olefinic C*H*), 3.90 (br, 1H) (cage C*H*), 3.57 (m, 1H) (BC*H*). $^{13}C\{^1H\}$ NMR (100 MHz, $CDCl_3$): δ 141.6, 139.0, 137.3, 132.6, 84.8, 71.9 (cage *C*), the bridged *C* connected to the cage B atom and another cage *C* were not observed. $^{11}B\{^1H\}$ NMR (128 MHz, $CDCl_3$): δ −2.5 (1B), −4.8 (1B), −5.3 (1B), −6.8 (1B), −9.0 (1B), −10.1 (1B), −12.3 (1B), −13.7 (1B), −14.3 (1B), −16.0 (1B). HRMS (EI) calcd for $C_8H_{15}^{11}B_8^{10}B_2{}^+Cl_2$: 288.1475. Found 288.1479.

1,3-(2-Trifluoro-2,5-cyclohexadiene-1,4-diyl)-*o*-carborane **V-3fa** or **V-3fb**: Yield: 32 %. White solid. ^1H NMR (400 MHz, $CDCl_3$): δ 7.28 (m, 1H), 6.92 (t, J = 6.8 Hz, 1H), 6.48 (t, J = 6.4 Hz, 1H) (olefinic C*H*), 4.27 (d, J = 6.8 Hz, 1H) (C*H*), 3.78 (t, J = 6.4 Hz, 1H) (BC*H*), 3.53 (br, 1H) (cage C*H*). $^{13}C\{^1H\}$ NMR (100 MHz, $CDCl_3$): δ 146.1, 142.7, 133.8, 122.5 (q, J_{C-F} = 267.9 Hz), 80.0 (cage *C*), 69.7 (cage *C*), 44.2, 34.1, the olefinic *C* connected to the C atom was not observed $^{11}B\{^1H\}$ NMR (128 MHz, $CDCl_3$): δ −3.1 (1B), −4.4 (1B), −6.3 (1B), −7.9 (1B), −9.5 (1B), −10.0 (1B), −12.2 (1B), −13.7 (1B), −14.6 (1B), −15.7 (1B). HRMS (EI) calcd for $C_9H_{15}^{11}B_8^{10}B_2{}^+F_3$ 288.2128. Found 288.2123.

1,3-(6-Trifluoro-2,5-cyclohexadiene-1,4-diyl)-*o*-carborane **V-3fa** or **V-3fb**: Yield: 20 %. White solid. ^1H NMR (400 MHz, $CDCl_3$): δ 7.31 (m, 1H), 6.88 (m, 1H), 6.36 (m, 1H) (olefinic C*H*), 4.22 (m, 1H) (C*H*), 3.81 (br, 1H) (cage C*H*), 3.78 (1H) (BC*H*) (overlapped). $^{13}C\{^1H\}$ NMR (100 MHz, $CDCl_3$): δ 146.0, 143.2, 132.0, 122.8 (q, J_{C-F} = 261.0 Hz) 80.0 (cage *C*), 70.3 (cage *C*), 44.2, 34.0, the olefinic *C* connected to the C atom was not observed $^{11}B\{^1H\}$ NMR (128 MHz, $CDCl_3$): δ −3.1 (1B), −4.3 (1B), −6.6 (1B), −8.3 (1B), −9.6 (1B), −10.3 (1B), −11.9 (1B), −14.4 (2B), −16.0 (1B). HRMS (EI) calcd for $C_9H_{15}^{11}B_8^{10}B_2{}^+F_3$ 288.2128. Found 288.2125.

3-(2-Methylphenyl)-*o*-carborane **V-4i**: Yield: 27 %. Colorless oil. ^1H NMR (400 MHz, $CDCl_3$): δ 7.37 (d, J = 7.6 Hz, 1H), 7.30 (t, J = 7.2 Hz, 1H), 7.20 (m, 2H) (aromatic C*H*), 3.83 (br, 2H) (cage C*H*), 2.67 (s, 3H) (CH_3). $^{13}C\{^1H\}$ NMR (100 MHz, $CDCl_3$): δ 143.9, 134.0, 131.6, 129.8, 125.8, 57.0 (cage *C*), 23.7, the aromatic *C* connected to the B atom was not observed $^{11}B\{^1H\}$ NMR (128 MHz, $CDCl_3$): δ −2.7 (2B), −5.4 (1B), −8.7 (1B), −11.8 (1B), −12.7 (2B), −13.6 (2B), −14.5 (1B). HRMS (EI) calcd for $C_9H_{16}^{11}B_8^{10}B_2{}^+$ 232.2250. Found 232.2247.

3-(2,3-Dimethylphenyl)-*o*-carborane **V-4j**: Yield: 39 %. Colorless crystals. ^1H NMR (400 MHz, $CDCl_3$): δ 7.19 (m, 2H), 7.08 (t, J = 7.6 Hz, 1H) (aromatic C*H*), 3.82 (br, 2H) (cage C*H*), 2.61 (s, 3H), 2.32 (s, 3H) (CH_3). $^{13}C\{^1H\}$ NMR (100 MHz, $CDCl_3$): δ 143.0, 137.9, 131.7, 131.4, 125.5, 57.3 (cage *C*), 21.3, 19.1, the aromatic *C* connected to the cage B atom was not observed $^{11}B\{^1H\}$ NMR (128 MHz, $CDCl_3$): δ −2.8 (2B), −5.2 (1B), −8.7 (1B), −12.1 (1B), −12.8 (2B), −13.8 (2B), −14.7 (1B). HRMS (EI) calcd for $C_{10}H_{20}^{11}B_8^{10}B_2{}^+$ 248.2568. Found 248.2560.

3-(2,6-Dimethylphenyl)-*o*-carborane **V-4ka**: ^{1}H NMR (400 MHz, $CDCl_3$): δ 7.25 (m, 2H), 7.17 (m, 1H) (aromatic C*H*), 4.10 (br, 2H) (cage C*H*), 2.68 (s, 6H) (CH_3). $^{13}C\{^1H\}$ NMR (100 MHz, $CDCl_3$): δ 144.2, 134.1, 129.2, 57.3 (cage *C*), 25.9, the aromatic *C* connected to the cage B atom was not observed.

3-(2,4-Dimethylphenyl)-*o*-carborane **V-4kb**: ^{1}H NMR (400 MHz, $CDCl_3$): δ 7.01 (m, 2H), 7.00 (d, J = 8.0 Hz, 1H) (aromatic C*H*), 3.81 (br, 2H) (cage C*H*), 2.63 (s, 3H), 2.33 (s, 3H) (CH_3). $^{13}C\{^1H\}$ NMR (100 MHz, $CDCl_3$): δ 143.8, 139.9, 132.5, 130.6, 126.6, 57.0 (cage *C*), 23.6, 21.2, the aromatic *C* connected to the cage B atom was not observed.

3-(2,5-Dimethylphenyl)-*o*-carborane **V-4l**: Yield: 41 %. Colorless oil. ^{1}H NMR (400 MHz, $CDCl_3$): δ 7.15 (s, 1H), 7.12 (m, 2H) (aromatic C*H*), 3.84 (br, 2H) (cage C*H*), 2.62 (s, 3H), 2.32 (s, 3H) (CH_3). $^{13}C\{^1H\}$ NMR (100 MHz, $CDCl_3$): δ 140.7, 135.0, 134.7, 131.6, 130.5, 57.0 (cage *C*), 23.2, 21.1, the aromatic *C* connected to the cage B atom was not observed $^{11}B\{^1H\}$ NMR (128 MHz, $CDCl_3$): δ −3.1 (2B), −5.7(1B), −9.0 (1B), −12.4 (1B), −13.1 (2B), −14.0 (2B), −14.9 (1B). HRMS (EI) calcd for $C_{10}H_{20}^{11}B_8^{10}B_2^+$ 248.2568. Found 248.2564.

3-(2,4,6-Trimethylphenyl)-*o*-carborane **V-4m**: Yield: 69 %. Colorless crystals. ^{1}H NMR (400 MHz, $CDCl_3$): δ 6.94(s, 2H), (aromatic C*H*), 4.01 (br, 2H) (cage C*H*), 2.67 (s, 6H), 2.31 (s, 3H) (CH_3). $^{13}C\{^1H\}$ NMR (100 MHz, $CDCl_3$): δ 144.2, 139.1, 131.6, 125.0, 56.3 (cage *C*), 25.8, 20.8. $^{11}B\{^1H\}$ NMR (128 MHz, $CDCl_3$): δ −3.2 (3B), −4.7 (1B), −7.6 (1B), −11.2 (1B), −12.8 (2B), −13.9 (2B). HRMS (EI) calcd for $C_{11}H_{22}^{11}B_8^{10}B_2^+$ 262.2725, Found 262.2726.

(1,3-Dimethyl-5-methylene-1,3-cyclohexadien-6-yl)-*o*-carborane **V-IM-4m**: ^{1}H NMR (400 MHz, $CDCl_3$): δ 5.65 (s, 1H), 5.58 (s, 1H), 4.73 (brs, 1H), 4.53 (brs, 1H) (olefinic C*H*), 3.56 (br, 1H), 3.49 (br, 1H) (cage C*H*), 3.04 (s, 1H) (C*H*), 1.98 (s, 3H), 1.74 (s, 3H) (CH_3).

1,3-(1,3,5-Hexatriene-1,6-diyl)-*o*-carborane **V-5a**: Colorless crystals. ^{1}H NMR (400 MHz, $CDCl_3$): δ 6.43 (d, J = 10.0 Hz, 1H), 6.01 (dd, J = 12.0, 3.6 Hz, 1H), 5.92 (m, 2H), 5.80 (d, J = 12.4 Hz, 1H), 5.69 (dd, J = 12.4, 4.0 Hz, 1H) (olefinic C*H*), 5.16 (br, 1H) (cage C*H*). $^{13}C\{^1H\}$ NMR (100 MHz, $CDCl_3$): δ 139.7, 132.7, 131.7, 126.3, 126.2, 73.2 (cage *C*), the olefinic *C* connected to the cage B atom and another cage *C* were not observed. $^{11}B\{^1H\}$ NMR (128 MHz, $CDCl_3$): δ −2.6 (1B), −4.8 (2B), −7.6 (1B), −9.8 (2B), −10.6 (1B), −11.2 (1B), −12.0 (1B), −12.8 (1B). Anal. Calcd for $C_8H_{16}B_{10}$: C, 43.61; H, 7.32. Found: C, 43.69; H, 7.38. HRMS (EI) calcd for $C_8H_{16}^{11}B_8^{10}B_2^+$ ($[M-H]^+$): 219.2175. Found 219.2176.

General Procedure for Pyrolysis of Cycloadducts

Pyrolysis of V-3a and V-5a A glass tube sealed with **V-3a** or **V-5a** (10.0 mg, 0.045 mmol) under vacuum was heated at 250 °C for 4 h or 6 h in a muffle furnace. After cooling to room temperature, the reaction mixture was examined by ^{1}H and ^{11}B NMR spectra and then subjected to flash column chromatography on silica gel (230–400 mesh) using *n*-hexane as eluent to give the desired product.

Procedure for Pyrolysis of Cycloadducts V-3a in Mesitylene Solution A glass tube sealed with a mesitylene (**V-2m**) solution (0.25 mL) or **V-3a** (10.0 mg, 0.045 mmol) was heated at 250 °C for 6 h in a muffle furnace. After cooling to room temperature and removal of solvent, the residue was examined by ^{1}H and ^{11}B NMR spectra and then subjected to flash column chromatography on silica gel (230–400 mesh) using *n*-hexane as eluent to give **V-4a** (9.8 mg, 98 %) as colorless crystals.

General Procedure for Synthesis of B(3)-substituted Carboranyl Alkenes. A Representative Procedure To a mixture of 3-Diazonium-*o*-carborane tetrafluoroborate (**VI-3**) (51.6 mg, 0.2 mmol) and LDA (21.4 mg, 0.2 mmol) was added alkene **VI-4** (1.0 mmol) in *n*-hexane (2 mL) at −30 °C under an atmosphere of dry argon. The resulting mixture was then stirred at room temperature for 10 min and quenched by wet *n*-hexane. After removal of the solvent in vacuo, the residue was examined by ^{1}H NMR spectrum and then subjected to flash column chromatography on silica gel (230–400 mesh) using *n*-hexane as eluent to give the desired product **VI-5**.

3-(2,3-Dimethyl-1-buten-3-yl)-*o*-carborane **VI-5a**: Colorless crystals. Yield: 92 %. ^{1}H NMR (400 MHz, $CDCl_3$): δ 4.81 (s, 1H), 4.65 (s, 1H) (olefinic C*H*), 3.32 (br, 2H) (cage C*H*), 1.83 (s, 3H), 1.32 (s, 6H) (CH_3). ^{13}C{^{1}H} NMR (100 MHz, $CDCl_3$): δ 153.7, 109.8, 56.2 (cage *C*), 28.8 22.5. ^{11}B{^{1}H} NMR (128 MHz, $CDCl_3$): δ −2.8 (2B), −4.8 (1B), −9.2 (1B), −12.9 (3B), −14.1 (3B). HRMS (EI) calcd for $C_8H^1_{22}B^{10}_8B_2{}^+$: 226.2731. Found 226.2727.

3-(4-Methyl-2-penten-1-yl)-*o*-carborane **VI-5b**: Colorless oil. It was isolated as an inseparable mixture of *E/Z* isomers (2.6:1). Yield: 87 %. Note: ^{1}H NMR was reported as observed, and distinction of isomers was made where possible. ^{1}H NMR (400 MHz, $CDCl_3$): δ 5.37–5.34 (major, m, 1.1H), 5.26–5.23 (minor, m, 0.46H) (olefinic C*H*), 3.36 (br, 2H) (cage C*H*), 2.55–2.53 (minor, m, 0.7H), 2.29–2.22 (major, m, 1.1H), 1.98–2.01 (minor, m, 1.0 H), 1.92–1.91 (major, m, 1.4H), 0.97–0.93 (m, 7.0H). ^{13}C{^{1}H} NMR (100 MHz, $CDCl_3$): δ 139.5, 124.1, 56.1 (cage *C*), 30.0, 22.9. ^{11}B{^{1}H} NMR (128 MHz, $CDCl_3$): δ −1.4 (2B), −3.6 (1B), −7.7 (1B), −11.4 (3B), −12.7 (3B).

3-(2-Ethyl-2-buten-1-yl)-*o*-carborane **VI-5c**: Colorless oil. It was isolated as an inseparable mixture of *E/Z* isomers (1:1.7). Yield: 97 %. Note: ^{1}H NMR was reported as observed, and distinction of isomers was made where possible. ^{1}H NMR (400 MHz, $CDCl_3$): δ 5.23 (major, q, J = 6.4 Hz, 0.7H), 5.05 (minor, q, J = 6.4 Hz, 0.3H) (olefinic C*H*), 3.34 (br, 2H) (cage C*H*), 2.21–1.96 (m, 4.7H), 1.58 (d, J = 6.8 Hz, 2.8H), 1.03–0.97 (m, 2.8H). ^{13}C{^{1}H} NMR (100 MHz, $CDCl_3$): δ 118.8, 116.9, 56.4 (cage *C*), 56.2 (cage *C*), 31.4, 24.2, 14.3, 12.8, 12.7. ^{11}B{^{1}H} NMR (128 MHz, $CDCl_3$): δ −2.9 (2B), −5.0 (1B), −9.2 (1B), −12.8 (3B), −14.2 (3B). NOESY spectrum confirmed the identity of the major isomer to be *Z*-isomer.

3-(1-Cyclohexen-1-ylmethyl)-*o*-carborane **VI-5d**: Colorless oil. Yield: 93 %. ^{1}H NMR (400 MHz, $CDCl_3$): δ 5.30 (s, 1H) (olefinic C*H*), 3.36 (br, 2H) (cage C*H*), 2.00 (d, J = 6.0 Hz, 4H), 1.88 (s, 2H), 1.63–1.59 (m, 4H). ^{13}C{^{1}H} NMR

(100 MHz, $CDCl_3$): δ 136.9, 121.8, 56.5 (cage *C*), 30.3, 25.4, 23.1, 22.4. $^{11}B\{^1H\}$ NMR (128 MHz, $CDCl_3$): δ −2.8 (2B), −4.8 (1B), −9.2 (1B), −12.9 (3B), −14.1 (3B). HRMS (EI) calcd for $C_9H_{22}^{11}B_8^{10}B_2^+$: 238.2837. Found 238.2839.

3-(2-Isopropyl-2-propen-3-yl)-*o*-carborane **VI-5ea**: Colorless oil. Yield: 63 %. ^1H NMR (400 MHz, $CDCl_3$): δ 4.76 (s, 1H), 4.56 (s, 1H) (olefinic C*H*), 3.37 (br, 2H) (cage C*H*), 2.26–2.19 (m, 1H), 2.05 (d, J = 8.0 Hz, 2H), 1.05 (d, J = 6.4 Hz, 6H). $^{13}C\{^1H\}$ NMR (100 MHz, $CDCl_3$): δ 107.0, 56.4 (cage *C*), 34.2, 29.8, another olefinic *C* was not observed. $^{11}B\{^1H\}$ NMR (128 MHz, $CDCl_3$): δ −1.4 (2B), −3.6 (1B), −7.7 (1B), −11.4 (3B), −12.7 (3B). HRMS (EI) calcd for $C_8H_{22}^{11}B_8^{10}B_2^+$: 226.2731. Found 226.2730.

3-(2,3-Dimethyl-2-buten-1-yl)-*o*-carborane **VI-5eb**: Colorless oil. Data were collect from the mixture of **VI-5ea** and **VI-5eb**. Yield: 31 %. ^1H NMR (400 MHz, $CDCl_3$): δ 3.34 (br, 2H) (cage C*H*), 1.91 (s, 2H), 1.70 (s, 3H), 1.65 (d, J = 3.2 Hz, 6H). $^{13}C\{^1H\}$ NMR (100 MHz, $CDCl_3$): δ 107.0, 56.1 (cage *C*), 31.1, 21.9, another olefinic *C* was not observed. $^{11}B\{^1H\}$ NMR (128 MHz, $CDCl_3$): δ −1.4 (2B), −3.6 (1B), −7.7 (1B), −11.4 (3B), −12.7 (3B).

3-(Cyclopenten-3-yl)-*o*-carborane **VI-5f**: Colorless oil. Yield: 83%. ^1H NMR (400 MHz, $CDCl_3$): δ 5.78 (d, J = 2.8 Hz, 1H), 5.66 (d, J = 3.2 Hz, 1H) (olefinic C*H*), 3.39 (br, 2H) (cage C*H*), 2.40–2.34 (m, 2H), 2.27–2.20 (m, 1H), 1.90–1.86 (m, 2H). $^{13}C\{^1H\}$ NMR (100 MHz, $CDCl_3$): δ 133.3, 131.0, 55.9 (cage *C*), 55.7 (cage *C*), 32.9, 29.8. $^{11}B\{^1H\}$ NMR (128 MHz, $CDCl_3$): δ −2.8 (2B), −4.8 (1B), −9.2 (1B), −12.9 (3B), −14.1 (3B). HRMS (EI) calcd for $C_7H_{18}^{11}B_8^{10}B_2^+$: 210.2306. Found 210.2303.

3-(Cyclohexen-3-yl)-*o*-carborane **VI-5g**: Colorless oil. Yield: 91 %. ^1H NMR (400 MHz, $CDCl_3$): δ 5.76-5.72 (m, 1H), 5.62 (d, J = 9.2 Hz, 1H) (olefinic C*H*), 3.45 (br, 2H) (cage C*H*), 1.97–1.93 (m, 3H), 1.78–1.73 (m, 2H), 1.61–1.57 (m, 2H). $^{13}C\{^1H\}$ NMR (100 MHz, $CDCl_3$): δ 128.8, 127.9, 55.2 (cage *C*), 28.1, 24.9, 22.1. $^{11}B\{^1H\}$ NMR (128 MHz, $CDCl_3$): δ −3.1 (3B), −9.2 (2B), −13.6 (3B), −14.1 (2B). HRMS (EI) calcd for $C_8H_{20}^{11}B_8^{10}B_2^+$: 224.2572. Found 224.2574.

3-(Cyclohexanon-2-yl)-*o*-carborane **VI-5h**: White solid. Yield: 98 % from **VI-4h**, 96 % from **VI-4i**. ^1H NMR (400 MHz, d_6-acetone): δ 6.94 (br, 1H) (OH), 4.37 (br, 2H) (cage C*H*), 2.52 (br, 4H), 1.95–1.93 (m, 4H). $^{13}C\{^1H\}$ NMR (100 MHz, d_6-acetone): δ 224.8, 172.7, 128.8, 86.8, 69.7 (cage *C*), 54.9, 51.5, 51.2. $^{11}B\{^1H\}$ NMR (128 MHz, d_6-acetone): δ −2.8 (1B), −5.0 (2B), −11.0 (1B), −13.8 (2B), −15.0 (2B), −16.3 (2B). HRMS (EI) calcd for $C_8H_{20}O^{11}B_8^{10}B_2^+$: 240.2566. Found 240.2561.

3-(Cyclohepten-3-yl)-*o*-carborane **VI-5j**: Yield: 83 %. ^1H NMR (400 MHz, $CDCl_3$): δ 5.86–5.79 (m, 1H), 5.66–5.62 (m, 1H) (olefinic C*H*), 3.46 (br, 2H) (cage C*H*), 2.20–2.14 (m, 3H), 2.00–1.95 (m, 1H), 1.86–1.83 (m, 1H), 1.70–1.51 (m, 2H), 1.50–1.47 (m, 2H). $^{13}C\{^1H\}$ NMR (100 MHz, $CDCl_3$): δ 133.8, 132.8, 56.3 (cage *C*), 55.8 (cage *C*), 32.2, 30.8, 28.5, 27.2. $^{11}B\{^1H\}$ NMR (128 MHz, $CDCl_3$): δ −2.9 (3B), −9.1 (2B), −13.3 (3B), −14.0 (2B). HRMS (EI) calcd for $C_9H_{22}^{11}B_8^{10}B_2^+$: 238.2838. Found 238.2842.

3-(Cycloocten-3-yl)-*o*-carborane **VI-5k**: Colorless oil. Yield: 76 %. ^{1}H NMR (400 MHz, $CDCl_3$): δ 5.77 (q, J = 6.0 Hz, 1H), 5.32 (t, J = 6.0 Hz, 1H) (olefinic C*H*), 3.49 (br, 2H) (cage C*H*), 2.39–2.36 (m, 3H), 2.33–2.20 (m, 2H), 1.91–1.87 (m, 2H), 1.80–1.70 (m, 2H), 0.88–0.86 (m, 2H). $^{13}C\{^{1}H\}$ NMR (100 MHz, $CDCl_3$): δ 132.0, 131.5, 56.3 (cage *C*), 35.4, 29.9, 27.7, 26.9, 26.1. $^{11}B\{^{1}H\}$ NMR (128 MHz, $CDCl_3$): δ −3.3 (3B), −9.6 (2B), −13.4 (2B), −14.3 (3B). HRMS (EI) calcd for $C_{10}H_{24}^{11}B_8^{10}B_2^{+}$: 252.3104. Found 252.3103.

3-(3-Methyl-2-methylene-3-buten-1-yl)-*o*-carborane **VI-5l**: Colorless oil. Yield: 92 %. ^{1}H NMR (400 MHz, $CDCl_3$): δ 5.16 (s, 1H), 5.13 (s, 1H), 5.07 (s, 1H), 4.90 (s, 1H) (olefinic C*H*), 3.34 (br, 2H) (cage C*H*), 2.36 (s, 2H), 1.92 (s, 3H). $^{13}C\{^{1}H\}$ NMR (100 MHz, $CDCl_3$): δ 146.1, 142.8, 114.9, 113.4, 56.2 (cage *C*), 21.0. $^{11}B\{^{1}H\}$ NMR (128 MHz, $CDCl_3$): δ −2.9 (2B), −5.3 (1B), −9.2 (1B), −12.7 (3B), −14.3 (3B). HRMS (EI) calcd for $C_8H_{20}^{11}B_8^{10}B_2^{+}$: 224.2572. Found 224.2570.

3-(3-Methyl-1,3-butadien-2-yl)-*o*-carborane **VI-5m**: Colorless oil. Yield: 92 %. ^{1}H NMR (400 MHz, $CDCl_3$): δ 5.60 (s, 2H), 4.91 (s, 1H), 4.78 (s, 1H) (olefinic C*H*), 3.53 (br, 2H) (cage C*H*), 1.92 (s, 3H). $^{13}C\{^{1}H\}$ NMR (100 MHz, $CDCl_3$): δ 147.1, 127.2, 112.8, 56.5 (cage *C*), 23.9. $^{11}B\{^{1}H\}$ NMR (128 MHz, $CDCl_3$): δ −1.9 (2B), −5.9 (1B), −9.1 (1B), −12.1 (4B), −13.8 (1B). HRMS (EI) calcd for $C_7H_{18}^{11}B_8^{10}B_2^{+}$: 210.2306. Found 210.2302.

General Procedure for Synthesis of B(3)-substituted Carboranyl Allenes. A Representative Procedure To a mixture of **VI-3** (51.6 mg, 0.2 mmol) and LDA (21.4 mg, 0.2 mmol) was added alkyne **VI-6** (1.0 mmol) in *n*-hexane (2 mL) at −30 °C under an atmosphere of dry argon. The resulting mixture was then stirred at room temperature for 10 min and quenched by wet *n*-hexane. After removal of the solvent in vacuo, the residue was examined by ^{1}H NMR spectrum and then subjected to flash column chromatography on silica gel (230–400 mesh) using *n*-hexane as eluent to give desired product **VI-7**.

3-(2,3-Hexadien-4-yl)-*o*-carborane **VI-7a**: Colorless oil. Yield: 92 %. ^{1}H NMR (400 MHz, $CDCl_3$): δ 5.10–5.05 (m, 1H) (allenic C*H*), 3.47 (br, 2H) (cage C*H*), 2.21–2.15 (m, 3H), 1.65 (d, J = 6.8 Hz, 2H), 1.07 (t, J = 7.2 Hz, 3H). $^{13}C\{^{1}H\}$ NMR (100 MHz, $CDCl_3$): δ 206.4, 85.1, 57.7 (cage *C*), 57.6 (cage *C*), 25.1, 14.1, 13.4. $^{11}B\{^{1}H\}$ NMR (128 MHz, $CDCl_3$): δ −1.4 (2B), −3.6 (1B), −7.7 (1B), −11.5 (3B), −12.8 (3B). HRMS (EI) calcd for $C_8H_{16}^{11}B_8^{10}B_2^{+}$: 224.2572. Found 224.2570.

3-(3,4-Octadien-5-yl)-*o*-carborane **VI-7b**: Colorless oil. Yield: 90 %. ^{1}H NMR (400 MHz, $CDCl_3$): δ 5.15–5.12 (m, 1H) (allenic C*H*), 3.46 (br, 2H) (cage C*H*), 2.17–2.11 (m, 2H), 2.03–1.96 (m, 2H), 1.55–1.48 (m, 2H), 1.00–0.94 (m, 6H). $^{13}C\{^{1}H\}$ NMR (100 MHz, $CDCl_3$): δ 205.5, 91.7, 57.8 (cage *C*), 57.5 (cage *C*), 34.2, 22.1, 21.6, 14.0, 14.0. $^{11}B\{^{1}H\}$ NMR (128 MHz, $CDCl_3$): δ −2.9 (2B), −6.4 (1B), −10.1 (1B), −13.2 (3B), −13.9 (2B), −14.7 (1B). HRMS (EI) calcd for $C_{10}H_{24}^{11}B_8^{10}B_2^{+}$: 252.3104. Found 252.3100.

3-(4,5-Decadien-6-yl)-*o*-carborane **VI-7c**: Colorless oil. Yield: 89 %. ^{1}H NMR (400 MHz, $CDCl_3$): δ 5.07–5.03 (m, 1H) (allenic C*H*), 3.47 (br, 2H) (cage C*H*), 2.17–2.14 (m, 2H), 1.98–1.93 (m, 2H), 1.50–1.43 (m, 3H), 1.41–1.33 (m, 3H),

0.95–0.90 (m, 6H). $^{13}C\{^1H\}$ NMR (100 MHz, $CDCl_3$): δ 205.9, 89.8, 57.8 (cage *C*), 57.6 (cage *C*), 31.7, 31.0, 30.8, 23.0, 22.5, 14.1, 14.0. $^{11}B\{^1H\}$ NMR (128 MHz, $CDCl_3$): δ −2.9 (2B), −6.3 (1B), −10.0 (1B), −13.0 (3B), −13.8 (2B), −14.5 (1B). HRMS (EI) calcd for $C_{12}H_{28}^{11}B_8^{10}B_2^+$: 280.3636. Found 280.3638.

3-(2,3-Pentadien-2-yl)-*o*-carborane **VI-7da**: Colorless oil. Yield: 43 %. ^1H NMR (400 MHz, $CDCl_3$): δ 4.96 (d, *J* = 3.2 Hz, 1H) (allenic C*H*), 3.47 (br, 2H) (cage C*H*), 2.22–2.18 (m, 3H), 1.87 (d, *J* = 2.0 Hz, 3H). $^{13}C\{^1H\}$ NMR (100 MHz, $CDCl_3$): δ 207.1, 82.9, 57.6 (cage *C*), 24.6, 18.6. $^{11}B\{^1H\}$ NMR (128 MHz, $CDCl_3$): δ −2.9 (2B), −6.6 (1B), −10.0 (1B), −13.0 (3B), −13.8 (2B), −14.6 (1B). HRMS (EI) calcd for $C_9H_{18}^{11}B_8^{10}B_2^+$: 234.2521. Found 234.2518.

3-(1,2-Pentadien-3-yl)-*o*-carborane **VI-7db**: Colorless oil. Data were collect from the mixture of **VI-7da** and **VI-7db**. Yield: 52 %. ^1H NMR (400 MHz, $CDCl_3$): δ 4.64 (d, *J* = 3.6 Hz, 2H) (allenic C*H*), 3.52 (br, 2H) (cage C*H*), 1.63 (d, *J* = 6.4 Hz, 2H), 1.11 (t, *J* = 7.2 Hz, 3H). $^{13}C\{^1H\}$ NMR (100 MHz, $CDCl_3$): δ 208.9, 73.9, 57.6 (cage *C*), 13.9, 13.3. $^{11}B\{^1H\}$ NMR (128 MHz, $CDCl_3$): δ −2.9 (2B), −6.6 (1B), −10.0 (1B), −13.0 (3B), −13.8 (2B), −14.6 (1B).

3-(2-Methyl-2,3-pentadien-4-yl)-*o*-carborane **VI-7ea**: Colorless oil. Data were collect from the mixture of **VI-7ea** and **VI-7eb**. Yield: 31 %. ^1H NMR (400 MHz, $CDCl_3$): δ 3.41 (br, 2H) (cage C*H*), 1.83 (s, 3H), 1.66 (s, 6H). $^{13}C\{^1H\}$ NMR (100 MHz, $CDCl_3$): δ 205.1, 92.3, 57.9 (cage *C*), 30.4, 20.2. $^{11}B\{^1H\}$ NMR (128 MHz, $CDCl_3$): δ −2.9 (2B), −6.6 (1B), −10.1 (1B), −13.0 (3B), −13.7 (2B), −14.8 (1B).

3-(4-Methyl-1,2-pentadien-3-yl)-*o*-carborane **VI-7eb**: Colorless oil. Yield: 63 %. ^1H NMR (400 MHz, $CDCl_3$): δ 4.64 (s, 2H) (allenic C*H*), 3.50 (br, 2H) (cage C*H*), 2.48-2.45 (m, 1H), 1.14 (d, *J* = 6.8 Hz, 6H). $^{13}C\{^1H\}$ NMR (100 MHz, $CDCl_3$): δ 208.2, 74.5, 57.9 (cage *C*), 30.4, 23.0. $^{11}B\{^1H\}$ NMR (128 MHz, $CDCl_3$): δ −2.9 (2B), −6.6 (1B), −10.1 (1B), −13.0 (3B), −13.7 (2B), −14.8 (1B). HRMS (EI) calcd for $C_8H_{20}^{11}B_8^{10}B_2^+$: 224.2572. Found 224.2576.

3-(3-Trimethylsilyl-1,2-pentadien-3-yl)-*o*-carborane **VI-7f**: Colorless oil. Yield: 87 %. ^1H NMR (400 MHz, $CDCl_3$): δ 4.20 (s, 2H) (allenic C*H*), 3.55 (br, 2H) (cage C*H*), 0.24 (s, 9H) (Si(CH_3)$_3$). $^{13}C\{^1H\}$ NMR (100 MHz, $CDCl_3$): δ 211.7, 64.0, 57.9 (cage *C*), −0.5. $^{11}B\{^1H\}$ NMR (128 MHz, $CDCl_3$): δ −1.9 (2B), −5.9 (1B), −9.1 (1B), −12.1 (3B), −13.8 (3B). HRMS (EI) calcd for $C_6H_{21}^{11}B_8^{10}B_2N^+$: 254.3585. Found 254.3583.

3-(1-Trimethylsilyl-1,2-pentadien-1-yl)-*o*-carborane **VI-7g**: Colorless oil. Yield: 83 %. ^1H NMR (400 MHz, $CDCl_3$): δ 4.77 (t, *J* = 6.0 Hz, 1H) (allenic C*H*), 3.48 (br, 2H) (cage C*H*), 2.01–1.93 (m, 2H), 0.97 (t, *J* = 7.2 Hz, 3H), 0.22 (s, 9H) (Si(CH_3)$_3$). $^{13}C\{^1H\}$ NMR (100 MHz, $CDCl_3$): δ 220.0, 82.8, 58.1 (cage *C*), 57.9 (cage *C*), 20.4, 14.2, −0.3. $^{11}B\{^1H\}$ NMR (128 MHz, $CDCl_3$): δ −2.9 (2B), −6.4 (1B), −10.0 (1B), −13.1 (3B), −14.6 (3B). HRMS (EI) calcd for $C_{10}H_{24}^{11}B_8^{10}B_2Si^+$: 280.3958. Found 280.3958.

3-(1-Trimethylsilyl-1,2-hexadien-1-yl)-*o*-carborane **VI-7h**: Colorless oil. Yield: 78 %. ^1H NMR (400 MHz, $CDCl_3$): δ 4.70 (t, *J* = 6.0 Hz, 1H) (allenic C*H*), 3.49 (br, 1H) (cage C*H*), 3.46 (br, 1H) (cage C*H*), 1.96–1.90 (m, 2H), 1.36

(q, J = 7.6 Hz, 2H), 0.93 (t, J = 7.6 Hz, 3H), 0.22 (s, 9H) (Si(CH_3)$_3$). $^{13}C\{^1H\}$ NMR (100 MHz, $CDCl_3$): δ 210.8, 80.9, 58.1 (cage C), 29.6, 23.5, 14.1, −0.3. $^{11}B\{^1H\}$ NMR (128 MHz, $CDCl_3$): δ −2.9 (2B), −6.4 (1B), −10.0 (1B), −13.2 (3B), −14.6 (3B). HRMS (EI) calcd for $C_8H^1_{22}B^{10}_8B_2{}^+$: 294.4225. Found 294.4229.

3 (4,4 Dimethylsilyl 1,2 pentadien-3-yl)-o-carborane **VI-7i**: Colorless oil. Yield: 72 %. ^1H NMR (400 MHz, $CDCl_3$): δ 4.60 (s, 2H) (allenic CH), 3.49 (br, 2H) (cage CH), 1.25 (s, 9H) (C(CH_3)$_3$). $^{13}C\{^1H\}$ NMR (100 MHz, $CDCl_3$): δ 208.5, 98.9, 59.0 (cage C), 30.4, 29.1. $^{11}B\{^1H\}$ NMR (128 MHz, $CDCl_3$): δ −2.9 (2B), −6.3 (1B), −10.0 (1B), −13.0 (3B), −13.8 (2B), −14.5 (1B). HRMS (EI) calcd for $C_9H^{11}_{22}B^{10}_8B_2{}^+$: 238.2837. Found 238.2839.

3-(3,4-Decadien-6-yn-5-yl)-o-carborane **VI-7j**: Colorless oil. Yield: 85 %. ^1H NMR (400 MHz, $CDCl_3$): δ 5.44 (t, J = 6.0 Hz, 1H) (allenic CH), 3.62 (br, 2H) (cage CH), 2.30–2.27 (m, 2H), 2.14–2.07 (m, 2H), 1.57–1.52 (m, 2H), 1.07 (t, J = 7.2 Hz, 3H), 0.98 (t, J = 7.2 Hz, 3H). $^{13}C\{^1H\}$ NMR (100 MHz, $CDCl_3$): δ 216.0, 93.1, 92.5, 56.5 (cage C), 29.9, 22.5, 21.7, 21.3, 13.7, 13.5. $^{11}B\{^1H\}$ NMR (128 MHz, $CDCl_3$): δ −2.9 (2B), −5.0 (1B), −9.2 (1B), −12.8 (3B), −14.2 (3B). HRMS (EI) calcd for $C_{12}H^{11}_{24}B^{10}_8B_2{}^+$: 276.2882. Found 276.2883.

3-(4,5-Dodecadien-7-yn-6-yl)-o-carborane **VI-7k**: Colorless oil. Yield: 80 %. ^1H NMR (400 MHz, $CDCl_3$): δ 5.37 (t, J = 6.4 Hz, 1H) (allenic CH), 3.61 (br, 2H) (cage CH), 2.32–2.29 (m, 2H), 2.09–2.04 (m, 2H), 1.52–1.45 (m, 3H), 1.42–1.36 (m, 3H), 0.98–0.88 (m, 6H). $^{13}C\{^1H\}$ NMR (100 MHz, $CDCl_3$): δ 216.3, 93.2, 90.6, 74.9, 56.6 (cage C), 31.1, 22.5, 22.2, 19.4, 14.3, 13.9, 13.7. $^{11}B\{^1H\}$ NMR (128 MHz, $CDCl_3$): δ −2.8 (2B), −4.9 (1B), −9.2 (1B), −12.6 (3B), −14.1 (3B). HRMS (EI) calcd for $C_{14}H^{11}_{28}B^{10}_8B_2{}^+$: 304.3414. Found 304.3413.

3-(2,7-Dimethyl-2,3-octadien-5-yn-4-yl)-o-carborane **VI-7l**: Colorless oil. Yield: 78 %. ^1H NMR (400 MHz, $CDCl_3$): δ 3.59 (br, 2H) (cage CH), 2.70–2.64 (m, 1H), 1.762 (s, 6H), 1.18 (d, J = 7.2 Hz, 6H). $^{13}C\{^1H\}$ NMR (100 MHz, $CDCl_3$): δ 207.2, 97.8, 88.1, 56.5 (cage C), 23.3, 22.6, 21.4, 19.5. $^{11}B\{^1H\}$ NMR (128 MHz, $CDCl_3$): δ −2.9 (2B), −5.0 (1B), −9.2 (1B), −12.7 (3B), −14.1 (3B). HRMS (EI) calcd for $C_{14}H^{11}_{28}B^{10}_8B_2{}^+$: 304.3414. Found 304.3417.

Reaction of 1,3-Dehydro-*o*-carborane with Cyclohexene-d_{10} under Standard Conditions To a mixture of **3** (51.6 mg, 0.2 mmol) and LDA (21.4 mg, 0.2 mmol) was added cyclohexene-d_{10} **VI-4g-d$_{10}$** (92.2 mg, 1.0 mmol) in n-hexane (2 mL) at −30 °C under an atmosphere of dry argon. The resulting mixture was then stirred at room temperature for 10 min and quenched by wet n-hexane. After removal of the solvent in vacuo, the residue was examined by ^1H NMR spectrum and then subjected to flash column chromatography on silica gel (230–400 mesh) using n-hexane as eluent to give the desired product **VI-5g-d$_{10}$** in 88% isolated yield. After ^1H and ^2H NMR analyses, 96 % deuterium incorporation at cage C–H was obtained.

Reaction of 1,3-Dehydro-*o*-carborane with 2-Methyl-1-hexen-3-yne (VI-8) under Standard Conditions To a mixture of **VI-3** (51.6 mg, 0.2 mmol) and LDA (21.4 mg, 0.2 mmol) was added 2-methyl-1-hexen-3-yne (**VI-8**) (94.2 mg, 1.0 mmol) in n-hexane (1.0 mL) at −30 °C under an atmosphere of dry argon.

The resulting mixture was then stirred at room temperature for 10 min and quenched by wet *n*-hexane. After removal of the solvent in vacuo, the residue was examined by ^{1}H NMR spectrum and then subjected to flash column chromatography on silica gel (230–400 mesh) using *n*-hexane as eluent to give desired product **VI-9a** and **VI-9b** in 91 % isolated yield with a ratio of **VI-9a/VI-9b** = 1.2/1.

3-(2-Methyleneyl-hexayn-1-yl)-*o*-carborane **VI-9a**: Colorless oil. Yield: 41 %. ^{1}H NMR (400 MHz, $CDCl_3$): δ 5.27 (s, 1H), 5.12 (s, 1H) (olefinic C*H*), 3.72 (br, 2H) (cage C*H*), 2.31–2.26 (m, 2H), 1.88 (s, 2H), 1.19–1.12 (m, 3H). $^{13}C\{^{1}H\}$ NMR (100 MHz, $CDCl_3$): δ 140.1, 120.7, 92.7, 80.9, 56.2 (cage *C*), 23.7, 13.8. $^{11}B\{^{1}H\}$ NMR (128 MHz, $CDCl_3$): δ −2.8 (2B), −4.9 (1B), −9.2 (1B), −12.8 (3B), −14.2 (3B). HRMS (EI) calcd for $C_9H_{20}^{11}B_8^{10}B_2^{+}$: 236.2681. Found 236.2682.

3-(2-Methyl-1,3,4-hexatrien-3-yl)-*o*-carborane **VI-9b**: Colorless oil. Yield: 50 %. Data were collect from the mixture of **VI-9a** and **VI-9b**. ^{1}H NMR (400 MHz, $CDCl_3$): δ 5.52 (s, 2H) (olefinic C*H*), 5.25 (q, *J* = 6.8 Hz, 1H) (allenic C*H*), 5.09 (s, 1H) (olefinic C*H*), 3.54 (br, 2H) (cage C*H*), 2.19 (s, 3H), 1.70 (d, *J* = 6.8 Hz, 2H). $^{13}C\{^{1}H\}$ NMR (100 MHz, $CDCl_3$): δ 210.3, 129.1, 115.3, 85.7, 58.5 (cage *C*), 58.4 (cage *C*), 14.5, 13.0. $^{11}B\{^{1}H\}$ NMR (128 MHz, $CDCl_3$): δ −2.8 (2B), −4.9 (1B), −9.2 (1B), −12.8 (3B), −14.2 (3B).

Reaction of 1,3-Dehydro-*o*-carborane with 1,1,1-trifluoroundec-3-yne (VI-6m) under Standard Conditions To a mixture of **3** (51.6 mg, 0.2 mmol) and LDA (21.4 mg, 0.2 mmol) was added 1,1,1-trifluoroundec-3-yne (**VI-6m**) (206.3 mg, 1.0 mmol) in *n*-hexane (2 mL) at −30 °C under an atmosphere of dry argon. The resulting mixture was then stirred at room temperature for 10 min and quenched by wet *n*-hexane. After removal of the solvent in vacuo, the residue was examined by ^{1}H NMR spectrum and then subjected to flash column chromatography on silica gel (230–400 mesh) using *n*-hexane as eluent to give only product **VI-7ma** in 73 % isolate yield.

3-(1-Trifluoro-2,3-undecadien-4-yl)-*o*-carborane **VI-7ma**: Colorless oil. Yield: 73 %. ^{1}H NMR (400 MHz, $CDCl_3$): δ 5.43 (q, *J* = 2.8 Hz, 1H) (allenic C*H*), 3.49 (br, 2H) (cage C*H*), 2.99–2.86 (m, 1H), 2.33–2.27 (m, 1H), 1.37–1.23 (m, 10H), 0.89–0.87 (m, 3H). $^{13}C\{^{1}H\}$ NMR (100 MHz, $CDCl_3$): δ 209.7, 125.8 (q, $^{1}J_{C\text{–}F}$ = 212.0 Hz), 85.1 (q, $^{2}J_{C\text{–}F}$ = 39.0 Hz), 57.4 (cage *C*), 31.9, 29.2, 29.1, 28.4, 28.0, 22.8, 14.2. $^{11}B\{^{1}H\}$ NMR (128 MHz, $CDCl_3$): δ −2.8 (2B), −4.8 (1B), −9.2 (1B), −12.9 (3B), −14.1 (3B). ^{19}F NMR (376 MHz, $CDCl_3$): −59.1 (d, *J* = 3.8 Hz, 3F). HRMS (EI) calcd for $C_{13}H_{27}^{11}B_8^{10}B_2F_3^{+}$: 348.3510. Found 348.3512.

Representative Procedure for the Light-mediated B(3)-Arylation of *o*-Carborane A CH_3CN solution of 3-diazonium-o-carborane tetrafluoroborate **VII-1** (25.8 mg, 0.1 mmol) and arene **VII-2** (0.2, 0.5 or 1.0 mmol) was stirred in the presence of light (a xenon lamp of optical power of 50 W as light source) and Eosin Y (0.02 mmol) as photocataylst under an atmosphere of dry argon. The resulting mixture was stirred at room temperature for 2 h and quenched by wet *n*-hexane. After removal of solvents in vacuo, the residue was examined by

^{1}H NMR spectrum and then subjected to flash column chromatography on silica gel (230–400 mesh) using *n*-hexane/EtOAc as eluent to give the desired products **VII-3**.

3-(Thiophen-2-yl)-*o*-carborane **VII-3a**: Colorless crystals. Yield: 86 %. ^{1}H NMR (400 MHz, $CDCl_3$): δ 6.71 (s, 2H), 6.21 (d, J = 9.6 Hz, 1H) (aromatic C*H*), 2.65 (br, 2H) (cage C*H*). $^{13}C\{^1H\}$ NMR (100 MHz, $CDCl_3$): δ 132.9, 131.3, 124.7, 56.2 (cage *C*). $^{11}B\{^1H\}$ NMR (128 MHz, $CDCl_3$): δ −2.5 (2B), −8.9 (1B), −9.6 (1B), −11.4 (1B), −12.8 (2B), −13.5 (3B). HRMS (EI) calcd for $C_6H_{14}^{11}B_8^{10}B_2S^+$: 226.1817. Found 226.1815.

3-(2-Phenyl-thiophen-5-yl)-*o*-carborane **VII-3b**: White solid. Yield: 83 %. ^{1}H NMR (400 MHz, CD_2Cl_2): δ 7.62 (dd, J = 8.4, 1.6 Hz, 2H), 7.40 (m, 3H), 7.33 (m, 2H) (aromatic C*H*), 3.89 (br, 2H) (cage C*H*). $^{13}C\{^1H\}$ NMR (100 MHz, CD_2Cl_2): δ 135.9, 129.5, 129.4, 128.4, 126.6, 126.3, 124.9, 124.5, 58.1 (cage *C*). $^{11}B\{^1H\}$ NMR (128 MHz, CD_2Cl_2): δ −2.8 (2B), −8.7 (2B), −13.3 (3B), −13.7 (3B). HRMS (EI) calcd for $C_{12}H_{18}^{11}B_8^{10}B_2S^+$: 305.2438. Found 305.2440.

3-(2-Methoxy-thiophen-5-yl)-*o*-carborane **VII-3c**: White solid. Yield: 93 %. ^{1}H NMR (400 MHz, CD_2Cl_2): δ 6.56 (m, 1.6 Hz, 1H), 6.49 (d, J = 2.0 Hz, 1H) (aromatic C*H*), 3.83 (s, 3H) (CH_3), 3.61 (br, 2H) (cage C*H*). $^{13}C\{^1H\}$ NMR (100 MHz, CD_2Cl_2): δ 135.6, 126.3, 118.3, 57.3 (cage *C*), 40.5. $^{11}B\{^1H\}$ NMR (128 MHz, CD_2Cl_2): δ −2.2 (2B), −8.6 (1B), −9.3 (1B), −11.2 (1B), −12.5 (2B), −13.3 (3B). HRMS (EI) calcd for $C_7H_{16}^{11}B_8^{10}B_2OS^+$: 256.1737. Found 256.1735.

3-(2,5-Dimethyl-thiophen-5-yl)-*o*-carborane **VII-3d**: White solid. Yield: 88 %. ^{1}H NMR (400 MHz, $CDCl_3$): δ 6.54 (s, 1H) (aromatic C*H*), 3.70 (br, 2H) (cage C*H*), 2.61 (s, 3H), 2.39 (s, 3H) (CH_3). $^{13}C\{^1H\}$ NMR (100 MHz, $CDCl_3$): δ 145.4, 136.6, 130.1, 56.9 (cage *C*), 15.6, 14.8. $^{11}B\{^1H\}$ NMR (128 MHz, $CDCl_3$): δ −2.8 (2B), −7.1 (1B), −9.2 (1B), −12.2 (2B), −13.0 (2B), −13.8 (2B). HRMS (EI) calcd for $C_8H_{18}^{11}B_8^{10}B_2S^+$: 254.2131. Found 254.2120.

3-(3,4-Dibromo-thiophen-2-yl)-*o*-carborane **VII-3e**: Colorless crystals. Yield: 78 %. ^{1}H NMR (400 MHz, CD_2Cl_2): δ 7.61 (s, 1H) (aromatic C*H*), 4.30 (br, 2H) (cage C*H*). $^{13}C\{^1H\}$ NMR (100 MHz, CD_2Cl_2): δ 132.8, 129.4, 125.8, 56.6 (cage *C*). $^{11}B\{^1H\}$ NMR (128 MHz, CD_2Cl_2): δ −2.7 (2B), −8.6 (2B), −11.0 (1B), −12.5 (2B), −13.5 (3B). HRMS (EI) calcd for $C_8H_{22}^{11}B_8^{10}B_2SBr_2^+$: 383.9388. Found 383.9387.

3-(Furan-2-yl)-*o*-carborane **VII-3f**: White solid. Yield: 90 %. ^{1}H NMR (400 MHz, $CDCl_3$): δ 7.51 (d, J = 1.2 Hz, 1H), 6.84 (d, J = 3.2 Hz, 1H), 6.41 (dd, J = 3.2, 1.6 Hz, 1H) (aromatic C*H*), 3.81 (br, 2H) (cage C*H*). $^{13}C\{^1H\}$ NMR (100 MHz, $CDCl_3$): δ 145.7, 119.7, 110.0, 56.5 (cage *C*). $^{11}B\{^1H\}$ NMR (128 MHz, $CDCl_3$): δ −2.7 (2B), −9.1 (1B), −9.8 (1B), −11.6 (1B), −13.0 (2B), −13.7 (3B). HRMS (EI) calcd for $C_7H_{18}^{11}B_8^{10}B_2O^+$: 210.2045. Found 210.2043.

3-(2-Methyl-furan-5-yl)-*o*-carborane **VII-3g**: White solid. Yield: 81 %. ^{1}H NMR (400 MHz, $CDCl_3$): δ 7.71 (d, J = 3.2 Hz, 1H), 5.98 (t, J = 1.6 Hz, 1H) (aromatic C*H*), 3.79 (br, 2H) (cage C*H*), 2.29 (s, 3H) (CH_3). $^{13}C\{^1H\}$ NMR (100 MHz, $CDCl_3$): δ 107.9, 107.1, 106.3, 56.5 (cage *C*), 13.7. $^{11}B\{^1H\}$ NMR

(128 MHz, $CDCl_3$): δ −3.0 (2B), −9.5 (2B), −12.1 (1B), −13.2 (1B), −14.0 (4B). HRMS (EI) calcd for $C_7H_{16}^{11}B_8^{10}B_2O^+$: 224.2202. Found 224.2207.

3-(2,5-Dimethyl-furan-3-yl)-*o*-carborane **VII-3h**: White solid. Yield: 78 %. ^{1}H NMR (400 MHz, CD_2Cl_2): δ 5.83 (s, 1H) (aromatic C*H*), 3.69 (br, 2H) (cage C*H*), 2.40 (s, 3H), 2.22 (s, 3H) (CH_3). $^{13}C\{^1H\}$ NMR (100 MHz, CD_2Cl_2): δ 110.1, 109.6, 109.3, 57.3 (cage *C*), 13.7, 13.1. $^{11}B\{^1H\}$ NMR (128 MHz, CD_2Cl_2): δ −3.0 (2B), −9.5 (2B), −12.1 (1B), −13.2 (1B), −14.0 (4B). HRMS (EI) calcd for $C_8H_{18}^{11}B_8^{10}B_2O^+$: 238.1355. Found 238.1357.

3-(1-Methyl-pyrrol-2-yl)-*o*-carborane **VII-3i**: White solid. Yield: 92 %. ^{1}H NMR (400 MHz, $CDCl_3$): δ 6.77 (s, 1H), 6.22 (t, J = 2.8 Hz, 1H), 6.10 (t, J = 2.8 Hz, 1H) (aromatic C*H*), 3.93 (br, 2H) (cage C*H*), 3.73 (s, 3H), (CH_3). $^{13}C\{^1H\}$ NMR (100 MHz, $CDCl_3$): δ 128.0, 116.6, 107.8, 57.8 (cage *C*), 37.0. $^{11}B\{^1H\}$ NMR (128 MHz, $CDCl_3$): δ −2.7 (2B), −9.1 (1B), −9.8 (1B), −11.6 (1B), −13.0 (2B), −13.7 (3B). HRMS (EI) calcd for $C_7H_{17}^{11}B_8^{10}B_2N^+$: 223.1241. Found 223.1247.

3-(1-*tert*-Butyloxycarbonyl-pyrrol-2-yl)-*o*-carborane **VII-3j**: White solid. Yield: 85 %. ^{1}H NMR (400 MHz, CD_2Cl_2): δ 7.38 (d, J = 1.2 Hz, 1H), 7.04 (s, 1H), 6.25 (t, J = 3.2 Hz, 1H) (aromatic C*H*), 4.40 (br, 2H) (cage C*H*), 1.56 (s, 9H) ($C(CH_3)_3$). $^{13}C\{^1H\}$ NMR (100 MHz, CD_2Cl_2): δ 150.8, 126.0, 112.5, 105.5, 85.2, 56.9 (cage *C*), 28.0. $^{11}B\{^1H\}$ NMR (128 MHz, CD_2Cl_2): δ −4.1 (2B), −8.1 (1B) −9.1 (1B), −11.3 (2B), −11.8 (1B), −14.5 (3B). HRMS (EI) calcd for $C_{11}H_{23}^{11}B_8^{10}B_2O_2N^+$: 309.2135. Found 309.2139.

3-(1-Methyl-2-acetyl-pyrrol-5-yl)-*o*-carborane **VII-3k**: White solid. Yield: 81 %. ^{1}H NMR (400 MHz, CD_2Cl_2): δ 6.92 (d, J = 4.0 Hz, 1H), 6.23 (d, J = 4.0 Hz, 1H) (aromatic C*H*), 4.21(s, 3H) (CH_3), 3.88 (br, 2H) (cage C*H*), 2.42 (s, 3H) (CH_3). $^{13}C\{^1H\}$ NMR (100 MHz, CD_2Cl_2): δ 197.0, 119.2, 116.1, 96.5, 58.3 (cage *C*), 36.4, 28.2. $^{11}B\{^1H\}$ NMR (128 MHz, CD_2Cl_2): δ −2.4 (2B), −8.3 (2B) −10.7 (1B), −12.3 (2B), −13.2 (3B). HRMS (EI) calcd for $C_9H_{19}^{11}B_8^{10}B_2ON^+$: 265.1609. Found 265.1611.

3-(Benzo[*b*]thiophen-2-yl)-*o*-carborane **VII-3l**: White solid. Yield: 84 %. ^{1}H NMR (400 MHz, CD_2Cl_2): δ 8.22 (d, J = 8.0 Hz, 1H), 8.00 (s, 1H), 7.71 (d, J = 8.0 Hz, 1H), 7.63 (m, 2H) (aromatic C*H*), 3.92 (br, 2H) (cage C*H*). $^{13}C\{^1H\}$ NMR (100 MHz, CD_2Cl_2): δ 135.0, 132.1, 130.1, 127.9, 125.6, 123.2, 122.5, 57.6 (cage *C*). $^{11}B\{^1H\}$ NMR (128 MHz, CD_2Cl_2): δ −3.2 (2B), −4.2 (1B) −8.7 (1B), −13.2 (3B), −14.1 (3B). HRMS (EI) calcd for $C_{10}H_{16}^{11}B_8^{10}B_2S^+$: 276.1976. Found 276.1976.

3-(1-Methyl-indol-2-yl)-*o*-carborane **VII-3m**: White solid. Yield: 82 %. ^{1}H NMR (400 MHz, CD_2Cl_2): δ 7.56 (d, J = 8.0 Hz, 1H), 7.37 (d, J = 8.0 Hz, 1H), 7.25 (m, 1H), 7.08 (m, 1H), 6.59 (m, 1H) (aromatic C*H*), 4.06 (s, 3H) (CH_3), 3.92 (br, 2H) (cage C*H*). $^{13}C\{^1H\}$ NMR (100 MHz, CD_2Cl_2): δ 130.3, 123.1, 120.9, 120.2, 110.0, 109.5, 102.8, 58.7 (cage *C*), 32.7. $^{11}B\{^1H\}$ NMR (128 MHz, CD_2Cl_2): δ −2.5 (2B), −8.3 (1B) −9.4 (1B), −11.5 (1B), −12.5 (1B), −13.2 (2B), −14.1 (2B). HRMS (EI) calcd for $C_{11}H_{19}^{11}B_8^{10}B_2N^+$: 273.2521. Found 273.2522.

3-(1,3-Dimethyl-indol-2-yl)-*o*-carborane **VII-3n**: White solid. Yield: 76 %. ^{1}H NMR (400 MHz, CD_2Cl_2): δ 7.56 (d, J = 8.0 Hz, 1H), 7.33 (d, J = 8.4 Hz, 1H), 7.26 (m, 1H), 7.09 (m, 1H) (aromatic C*H*), 4.15 (br, 2H) (cage C*H*), 4.05 (s, 3H), 2.44 (s, 3H) (CH_3). ^{13}C{^{1}H} NMR (100 MHz, CD_2Cl_2): δ 128.0, 123.4, 121.6, 119.5, 119.1, 111.8, 109.7, 56.6 (cage *C*), 33.5, 11.3. ^{11}B{^{1}H} NMR (128 MHz, CD_2Cl_2): δ −3.2 (2B), −7.9 (1B) −9.1 (1B), −11.4 (1B), −13.0 (2B), −13.9 (3B). HRMS (EI) calcd for $C_{12}H_{21}^{11}B_8^{10}B_2N^+$: 287.2678. Found 287.2676.

3-(1-Methyl-2-phenyl-indol-3-yl)-*o*-carborane **VII-3o**: White solid. Yield: 69 %. ^{1}H NMR (400 MHz, CD_2Cl_2): δ 8.08 (d, J = 8.0 Hz, 1H), 7.61 (d, J = 7.2 Hz, 1H), 7.55 (d, J = 1.6 Hz, 2H), 7.39 (m, 3H), 7.24 (m, 2H) (aromatic C*H*), 3.46 (s, 3H) (CH_3), 3.24 (br, 2H) (cage C*H*). ^{13}C{^{1}H} NMR (100 MHz, CD_2Cl_2): δ 130.9, 130.1, 129.5, 129.3, 129.0, 123.1, 122.8, 122.5, 121.3, 120.8, 110.0, 104.1, 56.6 (cage *C*), 30.7. ^{11}B{^{1}H} NMR (128 MHz, CD_2Cl_2): δ −3.2 (2B), −4.2 (1B) −8.7 (1B), −13.2 (3B), −14.1 (3B). HRMS (EI) calcd for $C_{17}H_{23}^{11}B_8^{10}B_2N^+$: 349.2836. Found 349.2835.

3-(1,3-Benzothiazol-2-yl)-*o*-carborane **VII-3pa**: White solid. Yield: 44 %. ^{1}H NMR (400 MHz, CD_2Cl_2): δ 8.02 (d, J = 13.2, 8.0 Hz, 2H), 7.53 (t, J = 8.0 Hz, 1H), 7.46 (t, J = 7.6 Hz, 2H) (aromatic C*H*), 4.26 (br, 2H) (cage C*H*). ^{13}C{^{1}H} NMR (100 MHz, CD_2Cl_2): δ 147.8, 137.4, 126.9, 126.2, 123.6, 122.3, 58.3 (cage *C*). ^{11}B{^{1}H} NMR (128 MHz, CD_2Cl_2): δ −2.8 (2B), −9.7 (1B) −11.2 (1B), −13.2 (3B), −14.3 (3B). HRMS (EI) calcd for $C_9H_{15}^{11}B_8^{10}B_2NS^+$: 277.1928. Found 277.1929.

3-(1,3-Benzothiazol-7-yl)-*o*-carborane **VII-3pb**: White solid. Yield: 17 %. ^{1}H NMR (400 MHz, CD_2Cl_2): δ 9.00 (s, 1H), 8.08 (dd, J = 12.0, 7.6 Hz, 2H), 7.51 (t, J = 7.6 Hz, 1H) (aromatic C*H*), 4.88 (br, 2H) (cage C*H*). ^{13}C{^{1}H} NMR (100 MHz, CD_2Cl_2): δ 155.0, 135.1, 135.0, 129.8, 125.8, 123.7, 57.8 (cage *C*). ^{11}B{^{1}H} NMR (128 MHz, CD_2Cl_2): δ −3.8 (2B), −6.2 (1B) −9.7 (1B), −12.1 (3B), −13.9 (3B). HRMS (EI) calcd for $C_8H_{20}^{11}B_8^{10}B_2NS^+$: 277.1928. Found 277.1930.

3-(4-Dimethylaminophenyl)-*o*-carborane **VII-3q**: White solid. Yield: 77 %. ^{1}H NMR (400 MHz, $CDCl_3$): δ 7.45 (d, J = 8.4 Hz, 2H), 6.70 (d, J = 12.0, 8.4 Hz, 2H) (aromatic C*H*), 3.65 (br, 2H) (cage C*H*), 2.98 (s, 6H) (CH_3). ^{13}C{^{1}H} NMR (100 MHz, $CDCl_3$): δ 151.6, 134.3, 112.1, 56.8 (cage *C*), 40.3. ^{11}B{^{1}H} NMR (128 MHz, $CDCl_3$): δ −2.9 (2B), −4.1 (1B) −8.7 (1B), −13.3 (3B), −14.0 (3B). HRMS (EI) calcd for $C_{10}H_{21}^{11}B_8^{10}B_2N^+$: 263.1881. Found 263.1886.

3-Phenyl-*o*-carborane **VII-3r**: White solid. Yield: 33 %. ^{1}H NMR (400 MHz, $CDCl_3$): δ 8.01–7.31 (m, 5H) (aromatic C*H*), 3.71 (br, 2H) (cage C*H*). The chemical shifts are in accordance with reported data.[162a]

3-(2,5-Dimethylphenyl)-*o*-carborane **VII-3s**: Yield: 47 %. White solid. ^{1}H NMR (400 MHz, $CDCl_3$): δ 7.15 (s, 1H), 7.12 (m, 2H) (aromatic C*H*), 3.84 (br, 2H) (cage C*H*), 2.62 (s, 3H), 2.32 (s, 3H) (CH_3). ^{13}C{^{1}H} NMR (100 MHz, $CDCl_3$): δ 140.7, 135.0, 134.7, 131.6, 130.5, 57.0 (cage *C*), 23.2, 21.1, the aromatic *C* connected to the cage B atom was not observed ^{11}B{^{1}H} NMR (128 MHz, $CDCl_3$): δ −3.1 (2B), −5.7(1B), −9.0 (1B), −12.4 (1B), −13.1 (2B), −14.0 (2B), −14.9 (1B). HRMS (EI) calcd for $C_8H_{20}O^{11}B_8^{10}B_2^+$ 248.2568, Found 248.2564.

3-(2,3-Dimethylphenyl)-*o*-carborane **VII-3ta**: Yield: 17 %. White solid. ^{1}H NMR (400 MHz, $CDCl_3$): δ 7.19 (m, 2H), 7.08 (t, J = 7.6 Hz, 1H) (aromatic C*H*), 3.82 (br, 2H) (cage C*H*), 2.61 (s, 3H), 2.32 (s, 3H) (CH_3). $^{13}C\{^{1}H\}$ NMR (100 MHz, $CDCl_3$): δ 143.0, 137.9, 131.7, 131.4, 125.5, 57.3 (cage *C*), 21.3, 19.1, the aromatic *C* connected to the cage B atom was not observed $^{11}B\{^{1}H\}$ NMR (128 MHz, $CDCl_3$): δ −2.8 (2B), −5.2 (1B), −8.7 (1B), −12.1 (1B), −12.8 (2B), −13.8 (2B), −14.7 (1B). HRMS (EI) calcd for $C_{10}H_{20}^{11}B_8^{10}B_2^+$ 248.2568, Found 248.2560.

3-(3,4-Dimethylphenyl)-*o*-carborane **VII-3tb**: Yield: 35 %. ^{1}H NMR (400 MHz, $CDCl_3$): δ 7.01 (m, 2H), 7.00 (s, 1H) (aromatic C*H*), 3.81 (br, 2H) (cage C*H*), 2.63 (s, 3H), 2.33 (s, 3H) (CH_3). $^{13}C\{^{1}H\}$ NMR (100 MHz, $CDCl_3$): δ 143.8, 139.9, 132.5, 130.6, 126.6, 57.0 (cage *C*), 23.6, 21.2. $^{11}B\{^{1}H\}$ NMR (128 MHz, $CDCl_3$): δ −2.8 (2B), −5.2 (1B), −8.7 (1B), −12.1 (1B), −12.8 (2B), −13.8 (2B), −14.7 (1B). HRMS (EI) calcd for $C_{10}H_{20}^{11}B_8^{10}B_2^+$: 248.2568, Found 248.2559.

3-(2,4,6-Trimethylphenyl)-*o*-carborane **VII-3u**: Yield: 69 %. Colorless crystals. ^{1}H NMR (400 MHz, $CDCl_3$): δ 6.94 (s, 2H), (aromatic C*H*), 4.01 (br, 2H) (cage C*H*), 2.67 (s, 6H), 2.31 (s, 3H) (CH_3). $^{13}C\{^{1}H\}$ NMR (100 MHz, $CDCl_3$): δ 144.2, 139.1, 131.6, 125.0, 56.3 (cage *C*), 25.8, 20.8. $^{11}B\{^{1}H\}$ NMR (128 MHz, $CDCl_3$): δ −3.2 (3B), −4.7 (1B), −7.6 (1B), −11.2 (1B), −12.8 (2B), −13.9 (2B). HRMS (EI) calcd for $C_{11}H_{22}^{11}B_8^{10}B_2^+$ 262.2725, found 262.2726.

3-Ferrocenyl-*o*-carborane **VII-3v**: Yield: 100 %. Colorless crystals. ^{1}H NMR (400 MHz, CD_2Cl_2): δ 4.51 (m, 4H), 4.22 (s, 5H) (ferrocenyl C*H*), 3.68 (br, 2H) (cage C*H*). $^{13}C\{^{1}H\}$ NMR (100 MHz, CD_2Cl_2): δ 73.7, 80.8, 69.5, 58.7 (cage *C*). $^{11}B\{^{1}H\}$ NMR (128 MHz, CD_2Cl_2): δ −3.2 (2B), −4.2 (1B), −8.7 (1B), −13.2 (2B), −14.1 (3B). HRMS (EI) calcd for $C_{12}H_{20}^{11}B_8^{10}B_2Fe^+$: 328.1918, found 328.1917.

MIX
Papier aus verantwortungsvollen Quellen
Paper from responsible sources
FSC® C105338

If you have any concerns about our products,
you can contact us on
ProductSafety@springernature.com

In case Publisher is established outside the EU,
the EU authorized representative is:
Springer Nature Customer Service Center GmbH
Europaplatz 3, 69115 Heidelberg, Germany

Printed by Libri Plureos GmbH
in Hamburg, Germany